Land use planning and remote sensing

REMOTE SENSING OF EARTH RESOURCES AND ENVIRONMENT

El-Baz, F.: Deserts and arid lands. 1984. ISBN 90-247-2850-9

Lindgren, D.T.: Land use planning and remote sensing. 1985.
ISBN 90-247-3083-X

Land use planning and remote sensing

By

DAVID T. LINDGREN

Department of Geography
Dartmouth College
Hanover, NH, USA

1985 **MARTINUS NIJHOFF PUBLISHERS**
a member of the KLUWER ACADEMIC PUBLISHERS GROUP
DORDRECHT / BOSTON / LANCASTER

Distributors

for the United States and Canada: Kluwer Academic Publishers, 190 Old Derby Street, Hingham, MA 02043, USA
for the UK and Ireland: Kluwer Academic Publishers, MTP Press Limited, Falcon House, Queen Square, Lancaster LA1 1RN, England
for all other countries: Kluwer Academic Publishers Group, Distribution Center, P.O. Box 322, 3300 AH Dordrecht, The Netherlands

Library of Congress Cataloging in Publication Data

Lindgren, David T.
Land use planning and remote sensing.

(Remote sensing of earth resources and environment ; 2)
Bibliography: p.
1. Remote sensing. 2. Land use--Remote sensing.
I. Title. II. Series.
G70.4.L56 1985 621.36'78 84-20624
ISBN 90-247-3083-X

ISBN 90-247-3083-X (this volume)
ISBN 90-247-2851-7 (series)

Copyright

PRINTED IN THE NETHERLANDS

To Paca

Preface

The purpose of this book is to introduce land planners to the principles of remote sensing and to the applications remote sensing has to the land planning process. The potential applications to land planning are many and varied. For example, remote sensing techniques, and aerial photography in particular, can provide planners with an overview of their communities they can obtain in no other way. These same techniques can also provide planners with a whole variety of land resource data and have the capability of updating these data on a systematic basis. Maps, too, can be produced from a combination of remote sensing and cartographic techniques – engineering maps, topographic maps, property maps, and a host of other thematic maps. These maps and the photos from which they are made can be used by planners to explain proposed land use or zoning changes at public meetings. They may also be introduced as evidence in courts of law if later the results of these changes are contested by individual or groups of landowners.

Since land planning tends to be conducted at local levels, the discussion in this book focuses on the uses of aerial photography – the most effective tool for small area analysis. The discussion is also directed at those who are not regular users of remote sensing techniques. But land planning is conducted over large areas as well – states, groups of states, and even whole countries, so Landsat is included in the discussion. Because the techniques involved in using Landsat effectively have become so sophisticated, no attempt is made to make this book a primer on the Landsat system.

This book will also ignore those remote sensing techniques where the applicability to land planning has not been clearly demonstrated. Accordingly, thermal scanning and side-looking airborne radar have received little mention. Instead, emphasis has been placed upon the sensors acquiring information in the visible and near-infrared portions of the spectrum, i.e. conventional camera systems and Landsat. It is for this reason that the book would not be a suitable text for an introductory remote sensing course unless supplemented by other materials.

The book is organized into two sections. The first, which includes Chapters 1–6, is concerned primarily with the principles of remote sensing and in particular aerial photography. Chapters cover the principles of electromagnetic radiation, aerial cameras and films, the geometry of aerial photos, fundamentals of photo-interpretation, and how to acquire photos. A separate chapter is devoted to the Landsat system. The second section of the book, Chapters 7–11, is concerned specifically with applications to land planning. Among the topics discussed are remote sensing inputs to geographic information systems, land use inventory and change, farmland and wetlands preservation, site selection, zoning, litigation, and methods of estimating population. No attempt is made to systematically cover all aspects of land planning, only those where remote sensing techniques may make some contribution.

As a final point the author should not be interpreted as suggesting that remote sensing is a panacea for all data and analysis needs; it is not. Remote sensing techniques primarily serve to complement more traditional ground survey techniques, although there may well be occasions when for reasons of time, manpower, and/or money remote sensing may be preferred even to ground surveys. It is the objective of this author to demonstrate when and for what purposes it is appropriate to use remote sensing techniques.

Contents

Acknowledgments

I am indebted to a number of individuals who provided assistance in the preparation of this book. Victor Klemas and his colleagues at the Center for Remote Sensing, University of Delaware, reviewed the first draft of the manuscript and offered a number of helpful comments. I owe particular thanks to Anniken Kloster who drafted all the diagrams appearing in the book. Virginia Perry was responsible for inputting the manuscript to the word-processor and for generating the final copies. Thanks must also go to the many individuals, representing universities, govenment agencies, and private firms, who generously provided illustrations and information.

I am most grateful to the Faculty Research Committee of Dartmouth College for its financial support of this project. Were it not for this support completion of this manuscript would have been difficult.

1. Introduction to remote sensing

1.1 Introduction

Remote sensing refers to the variety of techniques that have been developed for the acquisition and analysis of information about the earth. This 'information' is typically in the form of electromagnetic radiation that has either been reflected or emitted from the earth's surface. Since no single instrument is capable of detecting all of this radiation, a number of different sensors have been developed, each of which acquires energy measurements in a discrete portion of the electromagnetic spectrum. The distances from which these measurements are made vary greatly – from a few feet to thousands of miles. This capability has required the development of a wide range of sensor platforms. These platforms run the gamut from cherry pickers and tethered balloons to specially-designed high-altitude aircraft and earth-orbiting satellites (Fig. 1). Remote sensing acquisition techniques have become extremely complex.

The term 'remote sensing' is itself relatively new and was introduced in the early 1960s to replace the traditional but more restrictive terms 'aerial photography' and 'aerial photointerpretation.' The latter by definition refers only to the acquisition and analysis of data acquired by the conventional photographic process. Remote sensing, on the other hand, has a broader meaning encompassing the acquisition and analysis of data from all portions of the electromagnetic spectrum including the visible. But because so much of the discussion in this book focuses on the visible and near-infrared portions of the spectrum, the terms 'remote sensing' and 'aerial photointerpretation' will be used almost interchangeably. The same is also true for the terms 'photography' and 'imagery.' Though imagery in a technical sense refers to products made from electromechanical scanners and other devices measuring radiation beyond the visible portion of the spectrum, for purposes of this text the term will be used synonymously with 'photography'.

A word should also be said about data analysis techniques. As techniques for data acquisition have become more sophisticated so too have those for data analysis. Not many years ago the interpretation and measurement of features on black-and-white aerial photos were the only techniques available. Though

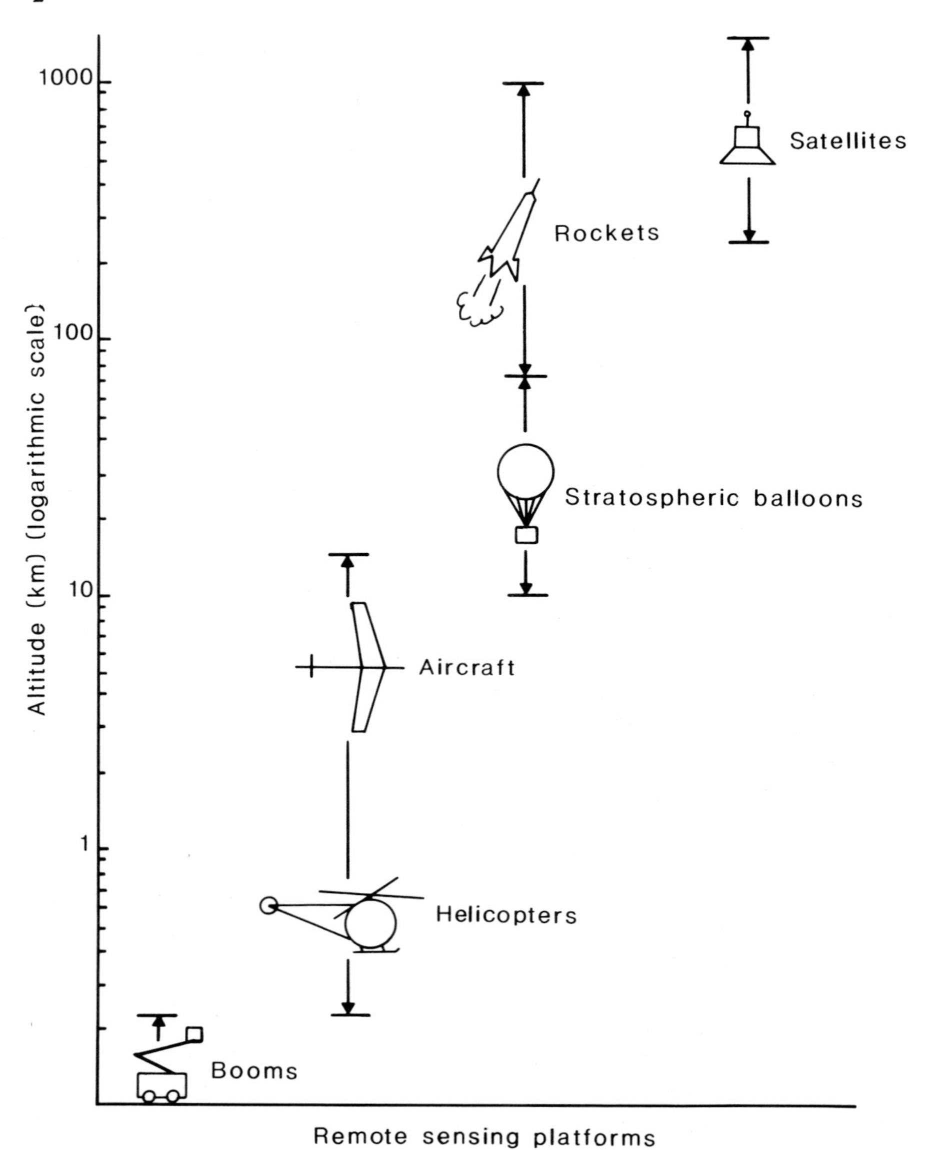

Fig. 1. Remote sensing platforms. Reproduced with permission from *Interpretation of Aerial Photographs*, Third Edition, copyright 1977, Burgess Publishing Company.

still important, these traditional photointerpretation techniques have been supplemented by a host of new techniques including digital analysis. Computers now make possible the use of larger data sets, more sophisticated analytical techniques, and a variety of graphic means of presenting analysis results. The objective of the data analysis process is to provide decisionmakers with the kinds of information that cannot be efficiently provided by more conventional methods.

The most appropriate place to begin a discussion of remote sensing techniques is with the primary data source – electromagnetic radiation. This chapter will briefly examine the nature of electromagnetic radiation as well as its various interactions with the atmosphere and the features comprising the earth's surface.

1.2 Electromagnetic Radiation

All materials at temperatures above absolute zero (0 K) produce electromagnetic energy. This energy is caused by the motions of the various charged particles that make up the atoms. Thus, in principle, any substance made up of atoms can produce electromagnetic energy; in practice, however, it is not that simple. The emission of energy depends primarily on the temperature of an object; the hotter the object, the greater is the emission of energy. However, it is also influenced by what is termed its emissivity. The emissivity of any object is a function of the chemical composition and physical state of that object.

To better understand the emission of electromagnetic energy from an object, the concept of 'black body' radiation is useful. A perfect black body reradiates all electromagnetic radiation that it receives. Most materials comprising the earth's surface, and including those produced by man, do not act as perfect black bodies but radiate less energy than the theoretical maximum for that temperature. If the theoretically maximum amount of electromagnetic radiation is taken to be 1, the actual emissivity of the various materials making up the earth's surface is less than 1.

Electromagnetic radiation is usually categorized either in terms of frequency or wavelength. Most scientists working within the field of remote sensing use wavelength to measure electromagnetic radiation and they do so principally because the interactions of electromagnetic radiation with the materials of the earth's surface differ as a function of wavelength. In other words, radio waves differ from light waves only in that the wavelengths of the former are very much longer. Yet, light waves do affect photographic film, while radio waves do not [1]. The arrangement of electromagnetic radiation according to wavelength (and frequency) is called the electromagnetic spectrum. While the spectrum is actually a continuum of wavelengths from microns to kilometers (Fig. 2), it has been arbitrarily divided into spectral regions on the basis of the sensors employed to record the various wavelengths. At present there is no single mechanism capable of acquiring information along the entire spectrum.

Most remote sensing systems operate in one of three regions of the electromagnetic spectrum – the optical or photographic, the infrared, and the micro-

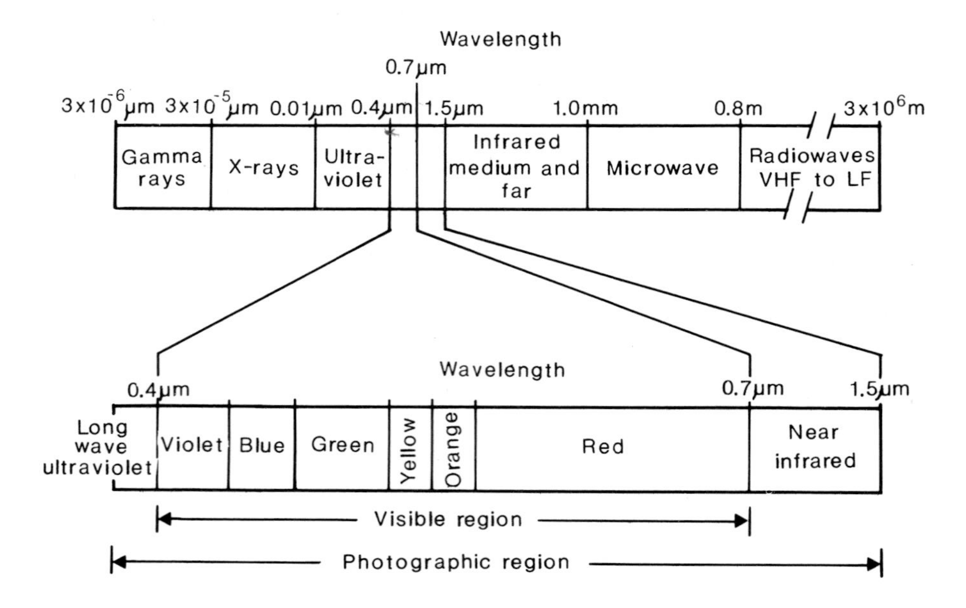

Fig. 2. Photographic region of electromagnetic spectrum. Reproduced with permission from *Remote Sensing for Planners*, copyright 1979, Center for Urban Policy Research, Rutgers – The State University of New Jersey.

wave. Each of these regions provide a unique type of data while requiring a specific acquisition technique.

1.3 Photographic wavelengths

The photographic wavelengths correspond closely to those perceived by the human eye. They are called 'photographic' because they cause chemical changes in film emulsions capable of producing images similar to those perceived by the human eye. Generally, this region is subdivided into the ultraviolet, the visible, and the near-infrared.

Ultraviolet wavelengths are relatively short (0.3 to 0.4 μm) and are not perceivable by the human eye. Furthermore, in order to record them, camera systems require special quartz lenses. These limitations have resulted in little use being made of the ultraviolet portion of the spectrum for remote sensing purposes and even less so for land use planning purposes.

The visible portion of the spectrum, that is those wavelengths directly perceivable by the human eye, extends from approximately 0.4 to 0.7 μm. When compared with the entire spectrum it is an extremely small portion, but because it corresponds to the wavelength perceived by the human eye it has traditionally been the most important portion. The human eye is capable of sensing narrow wavelengths within the visible that it perceives as different colors. Thus, the spectral region 0.4 to 0.5 μm the human eye perceives as blue, 0.5 to 0.6 μm as green, and 0.6 to 0.7 μm as red. The eye is not able, however, to isolate these wavelengths individually.

The near-infrared wavelengths are those that extend from 0.7 μm to about 1.1 μm. These wavelengths are not perceivable by the human eye. Nevertheless, they are similar to the visible wavelengths in that they penetrate conventional glass camera lenses and induce chemical changes in film emulsions. A common error is to assume near-infrared wavelengths are related to the sensation of heat. This is clearly not the case. Near-infrared wavelengths represent reflected, not emitted, electromagnetic radiation. They are known as near-infrared because they are 'near' the visible portion of the spectrum.

1.4 Infrared wavelengths

The infrared wavelengths represent a large spectral region extending theoretically from 0.7 μm to about 1000 μm. For this reason the region is commonly divided into three parts: the near-infrared which is reflected infrared energy and about which some comments have already been made; the middle-infrared (1.5 μm to 5.5 μm), a region composed of both reflected and emitted infrared wavelengths; and a far-infrared (5.5 μm to 1000 μm) composed of almost exclusively emitted infrared wavelengths. The infrared wavelengths longer than 1.5 μm respond quite differently than do the photographic wavelengths. They do not, for example, readily penetrate glass optics nor do they directly induce changes in the chemical composition of film emulsions. Rather optical-mechanical scanners have been developed to detect these wavelengths, which are recorded indirectly either on conventional films or digital tapes. One advantage of utilizing the emitted infrared wavelengths is that they may be acquired day or night. Unfortunately, this advantage is more than offset by the fact that infrared tend to be absorbed by atmospheric gases and water vapor. Data acquisition is, in fact, limited to two narrow spectral regions, or 'atmospheric windows' as they are commonly termed – 3 μm to 5 μm and 8 μm to 14 μm. While further mention will be made of the data acquired through these windows, it should be emphasized that little practical use has been made of the emitted infrared wavelengths for land use planning purposes.

1.5 Microwave wavelengths

Microwave wavelengths extend in length from one millimeter to almost a meter. They are, in essence, short radio waves and are acquired by a device similar to a

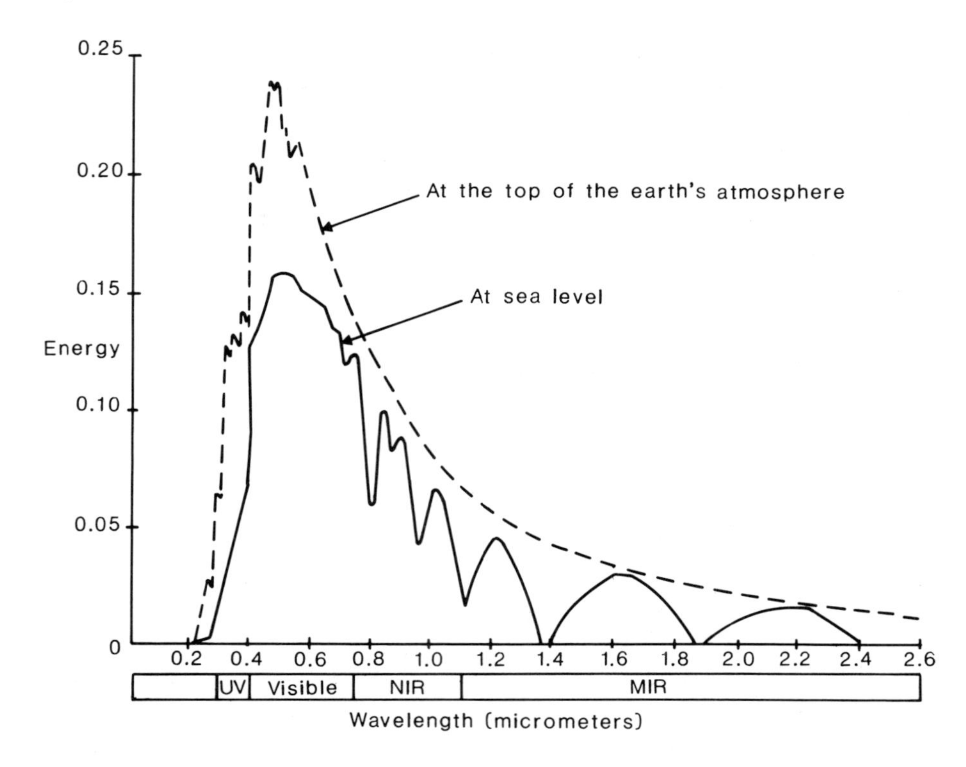

Fig. 3. Distribution of solar energy at outer edge of atmosphere and at sea level. Adapted from Department of the Army (1979).

radio receiver. Because microwave wavelengths are so long, at least in comparison with photographic and infrared wavelengths, they are not significantly attenuated by clouds. Thus, microwave systems can acquire data day and night and in almost any kind of weather. While systems for detecting microwave energy emitted by the earth's surface are available, most sensing is carried out with active systems such as radar. Radar systems generate their own radio waves that usually produce stronger returns or signals than the passive systems dependent upon natural microwave radiation. Microwave wavelengths have not been used extensively for land use purposes, although research conducted in this area has been promising.

1.6 Energy sources

The sun provides the major source of energy for passive remote sensing. It is not only extremely hot, with temperatures ranging from a few thousand degrees Kelvin to over a million degrees, but functions as a nearly perfect black body.

One result is that the sun emits radiation in all portions of the spectrum. However, as Figure 3 makes clear, solar radiation is heavily concentrated in the photographic and infrared spectral regions. About 45 percent of solar radiation consists of visible wavelengths, and another similar percentage of infrared wavelengths. This concentration of solar radiation in a relatively narrow portion of the spectrum unquestionably explains why human vision evolved in the spectral region it did.

1.7 Interaction of solar energy with atmosphere

If there were no atmosphere surrounding the earth, solar energy at all wavelengths would reach the earth's surface. As it happens, however, solar energy interacts with the various gases and particles comprising the atmosphere as it passes through. The result, illustrated in Figure 4, is to eliminate almost entirely certain wavelengths. In the visible portion of the spectrum, this loss of energy is due principally to scattering; in those portions of the spectrum on either side of the visible, adsorption plays the more important role.

Scattering is caused by solar insolation coming in contact with particles in the atmosphere. There are three principal types: Rayleigh scatter, Mie scatter, and non-selective scatter. Rayleigh scatter occurs at altitudes approximating 30,000 feet and is caused by gas molecules and other small particles. Because it primarily affects the shorter wavelengths, Rayleigh scatter is responsible for the bluish color of the sky on clear days. Mie scatter is caused by larger particles such as pollen, smoke, and dust which tend to be found in the lower layers of the atmosphere (under 15,000 feet). Where the atmosphere contains large concentrations of these particles, as frequently occurs over cities and industrial areas, Mie scatter causes the sky to take on a reddish appearance. Non-selective scatter is caused by the presence of large water droplets in the atmosphere – droplets with diameters several times those of incoming solar radiation. By definition non-selective scatter affects all wavelengths equally so concentrations of these droplets, i.e. fog and clouds, tend to appear white.

As a result of scattering, some solar energy is returned to space and lost forever. However, some of the scattered energy is directed earthward. This diffuse radiation, or skylight, has the effect of illuminating areas devoid of direct solar energy. As will be noted later, on aerial photos where shadows represent a near absence of information, scattering helps keep such information loss to a minimum.

Another type of energy loss is atmospheric absorption. Both carbon dioxide and water vapor are capable of directly absorbing infrared radiation. In fact, absorption is so great in certain regions of the infrared and microwave portions

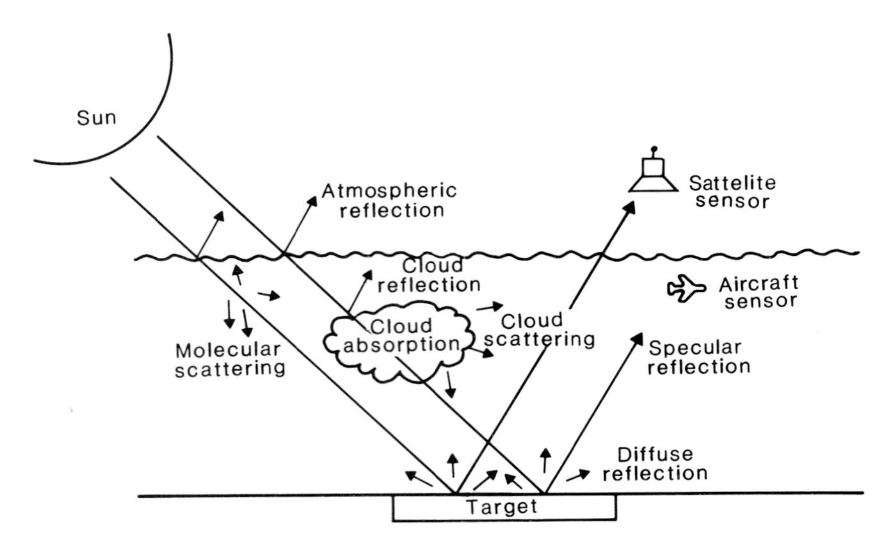

Fig. 4. Interactions of solar energy with earth's atmosphere and surface. Reproduced with permission from *Remote Sensing for Planners*, copyright 1979, Center for Urban Policy Research, Rutgers – The State University of New Jersey.

of the spectrum that they cannot be used to obtain remote sensor data from the earth's surface. Information gathering is limited to rather narrow bands or 'windows' through which radiation is able to pass largely unaffected by the atmosphere.

In summary, then, when the atmosphere is relatively clear, the combined effect of scattering and absorption may reduce the amount of solar energy reaching the earth by about 20 percent. However, this does not take into consideration the effect of cloud cover, which can also be effective in attenuating solar energy. Thus, under conditions of a heavy cloud layer, the combined reflection and absorption from clouds alone can account for a loss of from 35 to 80 percent of the incoming radiation and allow from 45 to 0 percent to reach the ground [2].

1.8 Interaction of solar energy with earth's surface

Solar energy falling upon the earth interacts with the various features of the earth's surface in several important ways – some is reflected, some transmitted, and some absorbed. Whether the energy is reflected, transmitted, or absorbed depends upon such factors as the wavelength, the angle at which the incident energy strikes the surface, and the composition of the surface materials. It is these differences that make it possible to distinguish between the multitude of earth features visible on aerial photos.

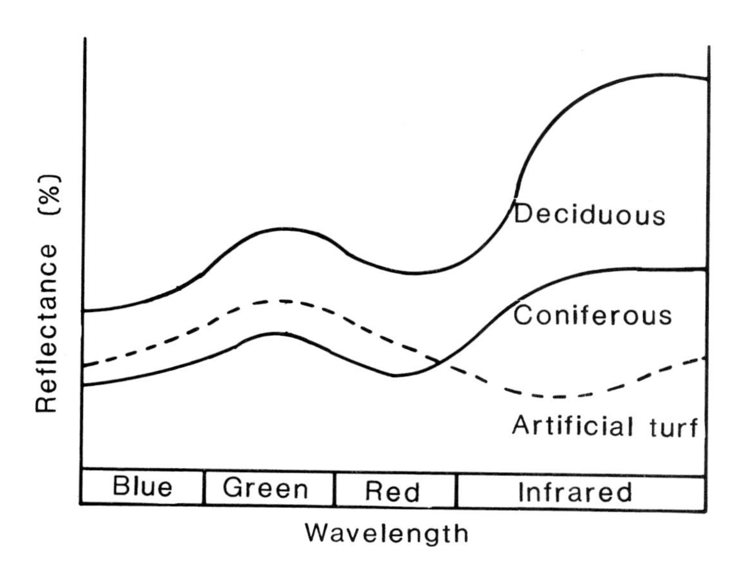

Fig. 5. Spectral reflectance curves for deciduous vegetation, coniferous vegetation, and artificial turf.

With solar energy so heavily concentrated in the reflected (photographic) portion of the spectrum, it is important to understand something of the reflectance properties of surface features. Generally, surfaces are considered as either specular or diffuse. *Specular* surfaces are perfectly smooth and energy striking them is reflected away at an angle equal to the angle of incidence (Fig. 4). No energy is returned in the direction of the source. Few surfaces are perfect reflectors, however, and most reflect energy in a more diffuse manner. Diffuse surfaces are rough and by definition the variations in surface height are larger than the wavelength of the energy falling upon them. Reflection is in all directions including back along the energy's path. The human eye perceives the energy reflected from diffuse surfaces as 'color'.

The percentage of each wavelength reflected from a surface can be computed and graphed to form *spectral reflectance curves*. These curves have been derived for many earth surface features (Fig. 5) though it should be emphasized that these are average curves compiled on the basis of a large sample of features. In fact, the spectral response of features varies greatly. For example, the reflection of solar energy from a leaf surface may change as the leaf grows to maturity, or as the leaf withers due to drought or disease. The same leaf species may differ between geographic locations because of differences in soils and moisture. And finally, any surface may vary in its reflectance characteristics as a result of atmospheric conditions, which may differ from day to day and place to place.

Thus, as we shall see, only a few of the earth's surface features can be consistently identified on the basis of spectral response alone.

It should be clear from even this brief discussion that the interactions between solar energy, the earth's atmosphere, and the features of the earth's surface are both numerous and complex. In many instances they are also incompletely understood. To effectively detect and record the energy reflected/emitted by the earth is a complicated task requiring a variety of sensing devices and involving a multitude of considerations. In the following section the procedures by which this energy is detected, recorded, and interpreted will be discussed.

2. Aerial cameras, filters, and films

2.1 Introduction

The visible wavelengths comprise the most utilized portion of the electromagnetic spectrum for remote sensing purposes. The reasons for this are several. First, the basic detection and recording device of the visible wavelengths, the aerial camera, has been available for over a century. The first aerial photos, from captive ballons, were acquired during the 1850's. Second, the spatial resolution of aerial photos (that is, the smallest objects clearly visible) tends to be considerably higher than that of nonphotographic systems, or at least of those systems available to the nonmilitary community. Third, the photos acquired by aerial cameras portray the earth's surface in a manner comparable to that perceived by the human eye. As a result, individuals can be trained quickly and easily to read and interpret aerial photos. And fourth, aerial camera systems tend to be less expensive to employ than nonphotographic systems. Thus, as Colwell has stated, 'Photos have been and will continue to be the basic remote sensing records; other remote sensing records will merely supplement them' [3].

This chapter will describe the more important characteristics of aerial camera systems, the filters and films most commonly used in conjunction with them, and the final products most often derived from them.

2.2 Aerial Cameras

The basic aerial camera is similar in principle to the hand-held camera. It is basically a lightproof box containing a specially designed film capable of recording electromagnetic radiation entering the box through a small aperture or opening. The box, or magazine, contains a lens that concentrates the electromagnetic radiation on the film, a shutter and stop that admits the correct amount of electromagnetic radiation, and a mechanism to periodically rotate the film. Filters may be placed over the lens to control the specific wavelengths allowed to enter the box. Many modifications have been made in the aerial camera, however. They are necessitated by the fact that the aerial camera must expose large numbers of photographs in rapid succession (aircraft speeds may exceed 1700 mph in the case of the SR-71 reconnaissance aircraft!), may have to acquire these photos through several miles of the earth's atmosphere, and must produce a final product embodying a high level of geometric fidelity.

There are many models of aerial cameras, but most would fall into one of four categories – single-lens, multi-lens, panoramic, and strip. Of these four types the single-lens camera is by far the most commonly used for land use planning purposes. For this reason it will be described in some detail; the other camera types will be discussed only briefly.

The *single-lens camera*, also referred to as the metric or mapping camera, is almost exclusively used for the preparation of maps utilizing photogrammetric techniques. It is a precision-built camera of great reliability and can be used with a variety of films, and produces the standard vertical photo in a 9 inch (230 mm) to the side format. The principal components of a single-lens camera are illustrated in Figure 6.

The typical single-lens camera can be broken down into two basic parts – the magazine and the lens cone. The magazine is a lightproof box that contains the film. While films vary, the most common format is 200 feet (60 m) long and 9.5 inch (240 mm) wide; approximately 250 exposures can be made from a single roll. A drive mechanism is responsible for drawing the film from the supply spool, across the film plate, and onto the take-up spool. The exposure is made as the film is drawn across the plate. To hold the film flat and thereby reduce any distortion on the photo, a suction is created behind the plate at the instant of exposure. The lens cone is also lightproof and contains the lens, filter (if any), shutter, and diaphragm. Its basic function is to direct and focus the desired electromagnetic radiation on the film. The distance between the center of the lens assembly and the film plate is called the camera's focal length. As will be discussed later, the longer the focal length, the larger the scale of the photos that can be taken. The most common focal lengths for aerial cameras are 152 mm, 210 mm, and 305 mm.

The lens assembly is composed of a series of high quality glass elements that serve to focus the radiation reflected from the earth's surface onto the photographic film. Because of the great height at which aerial photos are taken, the lenses on aerial cameras are generally focused at infinity and cannot be adjusted otherwise. On most aerial cameras the shutter and diaphragm are located between the front and rear lens assemblies. This shutter, termed a 'between the lens' shutter, controls the duration of the exposure (usually from 1/100 to 1/1000 sec.). The diaphragm forms the aperture through which the radiation passes. It can be varied to control the amount of radiation allowed through.

The camera body is that part of the camera assembly which attaches the lens cone and the magazine to the mounting on the aircraft. The body also houses

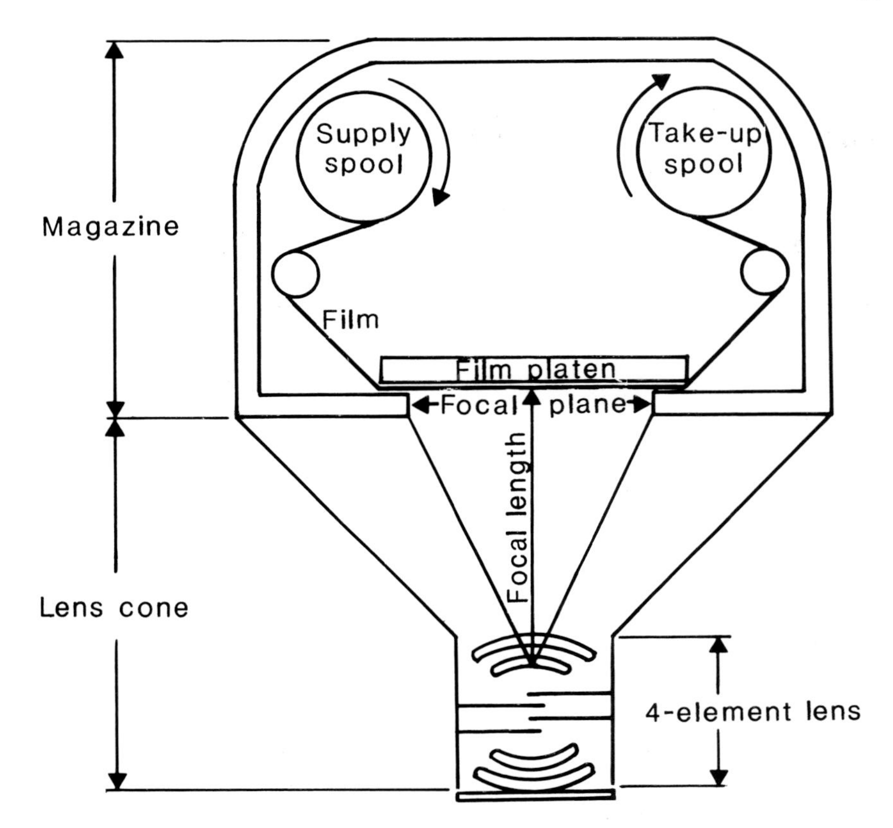

Fig. 6. Components of typical single-lens camera.

the altimeter, timepiece, and serial number counter. When the camera is shuttered the altitude, time, and serial number of that frame are automatically photographed and will appear on the margin of the finished photo.

A gimbal mounting attaches the camera to the aircraft. The mounting has two functions: it enables the camera to change its orientation in relation to the aircraft and it provides a shock absorber to protect the camera from aircraft vibrations.

While most of the aerial photos utilized by land use planners are acquired by single-lens cameras, on occasion it may be desirable to employ one of the other camera systems. The *multi-lens camera*, like the single-lens, takes high-resolution photos but additionally can take several of them simultaneously of the same area with each photo recording a different portion of the electromagnetic spectrum (visible and/or near-infrared). Cameras with as many as 16 lenses are available although three to four lens systems are the more common. The multi-lens camera would only be used if very detailed identifications were to be made. The *panoramic camera* produces high-resolution photography with wide swath

Fig. 7. Single-lens aerial mapping camera.

coverage. The lens scans the earth's surface through large angles perpendicular to the line of flight. Because of the great distortions in the photos, panoramic cameras are seldom used for mapping missions. However, panoramic cameras can provide coverage of large areas with a minimum of passes. The ***strip camera***, the last of the four types, acquires photography in a continuous vertical strip parallel to the line of flight. The film passes over a narrow slit opening in the camera at a rate equal to the speed of the aircraft. Strip photography is particularly useful for obtaining information on linear features such as pipelines, highways, railroads, and electrical transmission lines.

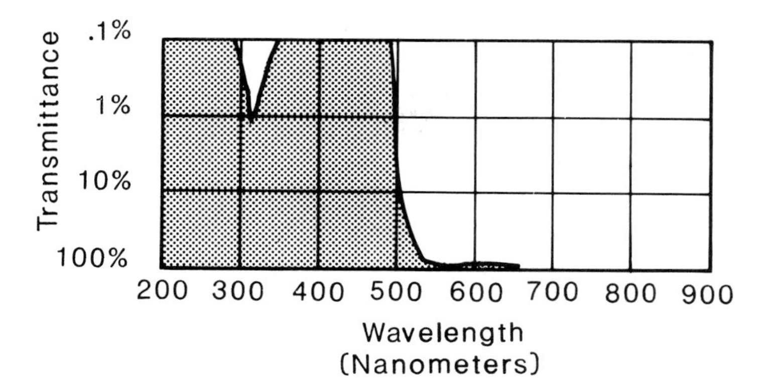

Fig. 8. Transmittance characteristics of medium yellow filter (Wratten 12).

2.3 Filters

Filters are transparent devices fitted over camera lenses to keep unwanted electromagnetic radiation from reaching the photographic film. Most aerial camera systems require some type of filter to eliminate the effects of atmospheric haze, which 'causes an overall bluish cast to color aerial photographs and lowers the contrast in aerial black-and-white negatives' [4].

The filters most commonly used in aerial camera systems are made of either gelatin or cellulose acetate into which dyes were dissolved during manufacture. The dyes, depending upon their color, act to absorb some wavelengths and transmit others. A *spectrophotometric absorption* or *transmittance curve* accompanies each filter to show exactly the percentage of absorption or transmittance of each wavelength. Figure 7 illustrates the absorption/transmittance characteristics of a medium yellow or 'minus blue' filter (Kodak Wratten 12). This filter absorbs all of the blue portion of the visible spectrum while allowing the green and red portions to pass through; a small amount of ultraviolet radiation is also transmitted. The effect of this filter is to reduce atmospheric haze and allow a sharper photograph to be taken.

Whenever a filter is used with any camera, a decrease occurs in the amount of radiation reaching the film. To compensate for this loss, the exposure time must be increased by an amount proportional to the radiation absorbed by the filter. The ratio of the increased exposure to the normal exposure is referred to as the *filter factor* and can usually be found on the data sheets accompanying each aerial film.

There are a great many filters available for use with aerial camera systems. These are short-wavelength filters that transmit the shorter wavelengths and absorb the longer wavelengths, long wavelength filters that transmit the longer

Table 1. Filters most commonly used with aerial films*

Wratten Filter no.	Description and Use
2A	Pale yellow; absorbs ultraviolet radiation below 450 nm. Used with black-and-white films to reduce haze at high altitudes.
2B	Pale yellow; absorbs ultraviolet radiation below 390 nm. More effective than 2A for reducing haze at high altitudes.
12	Deep yellow (minus blue); provides haze penetration. Used with color and black-and-white infrared films.
H-3, H-4, H-5	Red; absorbs all of blue and nearly all of green portion of spectrum. Used generally with infrared films although used with panchromatic films for forest surveys.
25	Light yellow; designed especially for color aerial photography to provide haze penetration. Since H series absorbs ultraviolet radiation they are not recommended for use at low altitudes (under 500 feet) or on clear days.
89B	Visually opaque; used with infrared-sensitive black-and-white films to confine exposure to far-red and infrared portions of the spectrum.

*Adapted from Eastman Kodak Company, *Kodak Filters for Scientific and Technical Uses*, 1970.

wavelengths and absorb the shorter wavelength, and band pass filters that absorb all wavelengths except for a narrow portion of the spectrum. There are also *antivignetting filters* used to produce photos of uniform sharpness and contrast. Antivignetting filters are designed to be highly transparent to reflected energy at their outer edges but progressively less so toward the center where reflected energy is strongest. They are often combined with haze-reduction filters. Table 1 provides information on some of the most widely used filters.

2.4 Films

Aerial films are ordinarily made of a cellulose acetate or polyester base that has been coated on one side with a light-sensitive emulsion and on the other with an antihalation backing that absorbs light and thereby prevents the formation of halos around bright objects. The emulsion layer is a gelatinous substance filled with crystals of silver halide. When reflected light is focused by the lens assembly onto the emulsion layer of the film, the halide crystals become sensitized in proportion to the amount of energy incident upon them. The image, thus captured, becomes visible after the film is processed.

Manufacturing technnology generally limits film sensitivities to a range of from $0.3\,\mu$ to $1.2\,\mu$. Within this range, however, the spectral response of emulsion layers may vary greatly. For example, emulsion layers may be sensitive to ultraviolet wavelengths, visible wavelengths, or near-infrared wavelengths. For land use planning purposes, only those films sensitive to the visible and near-infrared wavelengths are used.

In addition to their spectral sensitivity, aerial films also differ in terms of their 'speed'. A fast film, for example, can work effectively under conditions of less than bright sunlight because the silver halide crystals are very large. Unfortunately, the photos produced from fast films are apt to appear coarse and grainy. Slow speed films, on the other hand, contain very fine silver halide crystals that need bright sunlight for proper exposure. These fine crystals make possible photos of high resolution quality. On occasion, however, the long exposure time of slow speed films may cause some blurring of scene elements.

Most aerial films are capable of providing information useful to land use planners. Nevertheless, either because of cost considerations or some quality unique to a particular film type, certain films have proved more effective than others.

2.4.2. *Black-and-white films*. The two general types of black-and-white sensitized materials used by land use planners are panchromatic and infrared films. A *panchromatic* film is a negative material with a spectral sensitivity of $0.3\,\mu$ to $0.7\,\mu$, which approximates the sensitivity range of the human eye. For this reason alone, pan films, as panchromatic materials are often known, have long been preferred for aerial mapping and interpretation. They also have a wide exposure latitude, excellent resolving power, and provide prints with good tonal contrast. Tonal contrast is enhanced further by the use of a yellow filter that eliminates the ultraviolet and blue wavelengths. Last, but not least, pan films tend to be considerably less expensive than alternative films such as conventional color and color infrared.

One of the factors contributing to pan films' cheaper cost is the ease with which they can be processed. Even many amateur photographers process their own panchromatic films. Basically, once a panchromatic film has been exposed, it is removed from the camera and immersed in a developer solution where the sensitized silver halide crystals are changed into particles of black silver. Next, the print is immersed in diluted acetic acid to stop the action of the developer. The film is then placed into a fixer solution to remove the unaffected silver halide crystals from the emulsion, washed and dried. The end result of this process is a negative image of the original scene. Where light was reflected, the image is clear. To get the correct representation of the original scene, a positive print must be made from the negative. This is accomplished by passing light through the negative such that it exposes the emulsion on a sheet of printing paper. The print is then developed and processed in much the same manner as the film.

Images on panchromatic films are portrayed in various shades of gray.

Unfortunately, the human eye is quite limited in the number of gray tones it can differentiate. If the contrast in scene elements is not great, then photo identification can be difficult. Pan films are not, for example, effective for vegetation studies; such films have little sensitivity in the green portion of the spectrum and subtle color differences are lost. On the other hand, pan films are excellent for land use studies because there is great contrast between man-made features and the features of the natural landscape.

If greater detail in vegetation is necessary, *black-and-white infrared* may be the more appropriate film. Made only by Eastman Kodak Company, black-and-white infrared film is spectrally sensitive from 0.36 μ to 0.9 μ. This means, in fact, that the film is sensitive to ultraviolet, visible, and some near-infrared wavelengths. However, the untraviolet and blue wavelengths are eliminated through the use of a yellow filter, while on occasion a red filter may be employed so that only red and infrared wavelengths are allowed to expose the film. Because the shorter wavelengths are always filtered out, black-and-white infrared film has a natural haze penetration capability.

In a healthy state natural vegetation is highly infrared reflective; cultural features, on the other hand, are not. Thus, natural vegetation appears light gray on infrared film while cultural features appear much darker. The film was used as a camouflage-detecting device during World War II, that is, to differentiate natural vegetation from military targets painted to look like vegetation from several thousand feet up. Today, the film is used primarily for vegetation studies. Not only, for example, can hardwoods (lighter tones) be differentiated easily from softwoods (darker tones) but information can also be acquired on the health of vegetation. Under stress, either from disease or insect infestation, vegetation loses infrared reflectivity and should, in principle, appear darker on the film. Infrared film may also be useful for hydrologic studies because water tends to absorb infrared wavelengths. In wetland areas infrared film will clearly and easily differentiate water courses (black) from vegetation (light).

2.4.3. *Color films*. There are three types of color films available to land use planners – color positive, color negative, and color infrared. The advantages of color films over black-and-white are fairly obvious. The human eye can discriminate between an almost infinite number of different colors but only a few hundred gray tones. This capability makes the identification of landscape features on color films both more accurate and more rapid. On the other hand, color films are more expensive than black-and-white films; they also require more careful use. If exposure settings and filter combination are not correct, the resulting photos may have a very washed-out appearance. Furthermore, the processing of color films 'is much more complex than processing black-and-white films and as such requires a high degree of regulation each step of the way. All of the care which is taken to correctly expose the film may be completely lost if the development is not precisely controlled.' [5].

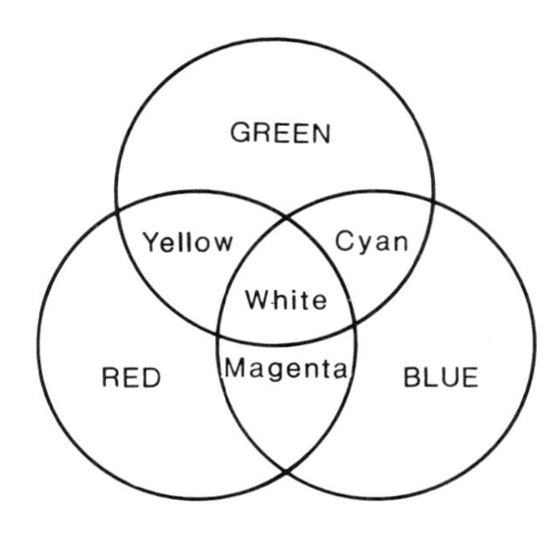

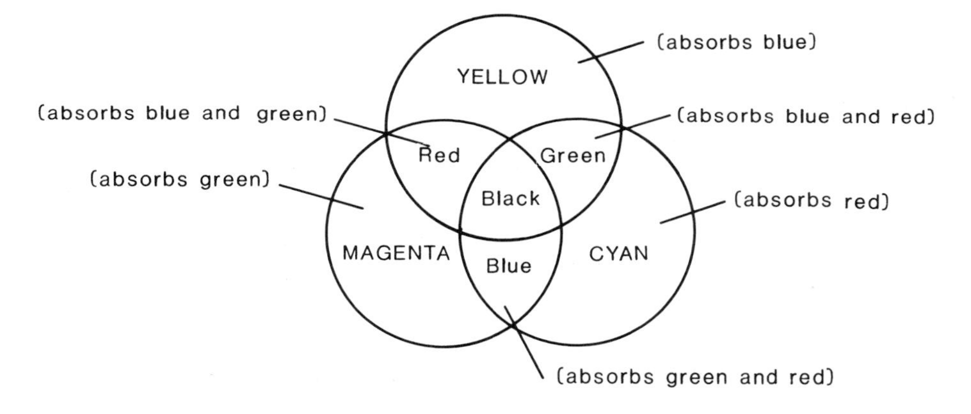

Fig. 9. Color mixing process. Adapted from Lillesand with permission.

The process by which photos are produced in color is based upon the fact that the colors perceived by the human eye can be duplicated by the proper mixing of three colors: blue, green, and red. These colors are known to photographers as primary colors or additives. Thus, for example, when blue and green light are superimposed the color cyan is produced, when green and red are superimposed the color yellow is produced, and when red and blue are superimposed magenta is produced. The superimposition of all three primary colors produces the sensation of white (see Fig. 9).

The colors cyan, yellow, and magenta are referred to as the secondary or subtractive colors because they require two of the primaries to produce them, or to put it another way, any one of the secondary colors will transmit two-thirds of the visible spectrum and absorb one-third. The secondaries are also known as the complementary colors. When the proper primary color, or complement, is added to one of the secondary colors white light is produced. Thus, cyan plus its complement red produces white as does yellow and its complement blue, and

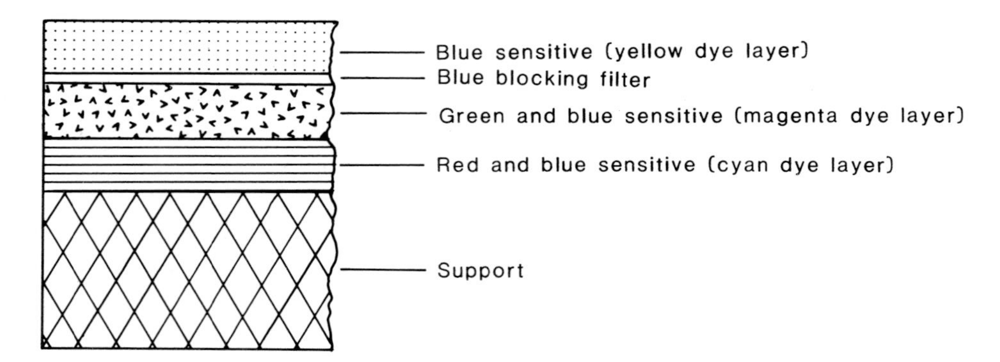

Fig. 10. Structure of color film. Reproduced with permission from *Remote Sensing and Image Interpretation*, copyright 1979, John Wiley and Sons.

magenta and its complement green. On occasion the secondary colors – cyan, yellow, and magenta – are also given the names – minus red, minus blue, and minus green.

Color films are composed of three emulsion layers each of which is sensitive to one of the primary colors (Fig. 10). The top layer is sensitive only to blue light; green and red light pass through this layer without effect. The middle emulsion is sensitive to green light and the bottom layer to red light. A yellow filter layer has been placed between the first and second layers to prevent any blue light from exposing the lower two layers. Each of the layers also includes a dye built in during manufacture. The dyes used are the complements of the primary colors to which the emulsion layers are sensitive. Thus, the blue sensitive layer contains a yellow dye, the green sensitive layer a magenta dye, and the red sensitive layer a cyan dye.

The processing of color films is a more complex procedure than the processing of black-and-white films and differs according to whether a film is color negative or color positive. In the case of *color negative*, the exposed film is placed into a developer solution which combines with a chemical substance (known as a coupler) in each layer to produce a colored compound or dye. The amount of dye produced is in direct proportion to the amount of black silver formed. Thus, the black silver in the blue-sensitive layer is replaced by yellow dye, in the green-sensitive layer by magenta dye, and in the red-sensitive layer by cyan dye. When the film has been properly washed and dried, there remains a negative which is the complement of the original colored scene.

A color print is produced from a color negative in much the same way as a print is produced from black-and-white film. White light is directed through the color negative and onto printing paper which in this instance is coated with a three-layer emulsion similar to that on the color film. Where yellow dye appears on the negative only green and red light will pass through to the printing paper. The green and red light thus sensitizes the green and red emulsion layers which when put into a developer solution will produce magenta and cyan dyes. On the finished print only blue light reflecting off the backing paper will be visible since green light will be absorbed by the magenta dye and red light by the cyan dye. The blue color appearing on the finished print should be comparable to the color of the object in the original scene.

The major advantage in using color negative as opposed to color positive films is the negative itself. From this negative a whole variety of products can be produced: color transparencies, color prints, black-and-white transparencies, and black-and-white prints. Because of this versatility, color negative films are among the most popular aerial films.

2.4.4. *Color positive* (or reversal) film differs from color negative film in that it may be processed to produce a positive transparency directly from the original film. Since overflights often involve hundreds of frames of photography, the savings in time and money from not having to produce a separate negative for each frame can be considerable. The disadvantage of such a film is obvious. If additional photographic products are needed at some later date, internegatives may have to be produced first, and the costs may be high. Contact prints may also be made but the quality is usually something less than the original.

Color positives are processed by first immersing the exposed film in a black-and-white developer which acts to produce a negative silver image in each of the three emulsion layers. The film is then reexposed to white light sensitizing the silver halides not exposed originally. This time the film is placed in a color developer causing the dyes to form in proportion to the silver developed. The salts originally exposed are washed away leaving only the dyes of the layers not originally exposed. The result is a color transparency. In the color negative and the reversal processes, the final product is actually three individual images superimposed on one another.

The spectral sensitivity of the three dye layers may be changed to respond to other photographic spectral regions including near-infrared. Such an infrared-sensitive color film was developed during World War II to complement the black-and-white infrared film developed earlier. Originally known as 'camouflage-detection film', the color version is more frequently referred to as 'color-infrared film' or 'false-color film'. The manufacturer's name for the film is Kodak Aerochrome Infrared, film number 2443. It is available only as a color positive or reversal.

The easiest way to describe color-infrared film is to compare it with conventional color film. A glance at Table 2 reveals several significant differences. First, while there are three emulsion layers in color-infrared film, they are sensitive to green (0.5–0.6 μ), red (0.6–0.7 μ), and near-infrared radiation (0.7–0.9 μ) rather than blue, green, and red as is the case with color film. Furthermore, while the same dyes are used in both types of films, they are used

Table 2. Spectral characteristics of color and color-infrared films

	Spectral Region			
	Blue	Green	Red	Infrared
Color film				
Sensitivity bands	Blue	Green	Red	
Corresponding colors of dye layers	Yellow	Magenta	Cyan	
Resulting colors on photos	Blue	Green	Red	
Color-infrared film				
Sensitivity bands with yellow filter		Green	Red	Infrared
Corresponding colors of dye layers		Yellow	Magenta	Cyan
Resulting colors on photos		Blue	Green	Red

for progressively longer wavelengths in the false color film. Also a yellow filter is used with false color films to eliminate the blue wavelengths to which all three emulsion layers are sensitive. The result is to allow only green, red and near-infrared radiation to reach the film. When color infrared film is properly processed, objects which reflect primarily green wavelengths will appear as blue-green, objects which reflect exclusively red wavelengths will appear green, and objects which are highly infrared reflective will appear red. These objects, then, will not appear as we are accustomed to seeing them and hence the term 'false-color film.'

Like its predecessor, color-infrared film was developed for the purpose of detecting military and industrial facilities that had been painted to look like natural vegetation from the air. Because healthy vegetation is highly infrared reflective it appears reddish in color-infrared photos; objects painted green, on the other hand, will appear blue on the film and can be distinguished immediately from the red of natural vegetation. Figure 5 illustrates this well. While natural vegetation and artificial turf have similar spectral sensitivities in the visible portion of the spectrum, they differ widely in the near-infrared portion.

It is the sensitivity of color-infrared film to changes in the health of vegetation that has been responsible for the interest shown in its use. Healthy vegetation is highly infrared reflective. Since the dye responses are inversely proportional to exposure of each layer, little of the cyan dye linked to the infrared-sensitive layer is produced. Instead, it is the magenta and yellow dyes (linked to the red- and green-sensitive layers) which are produced and in combination they cause healthy vegetation to appear reddish. When vegetation begins to lose its vigor, either through disease, insect infestation, or soil infertility, it loses infrared reflectance. This causes more cyan to show and proportionately less of the yellow and magenta.

When loss of plant vigor occurs to broad-leaved plants the affected vegetation appears dark red to black on color-infrared film. However, if plant stress is due to moisture loss infrared reflectance is reduced while the visible portion of the spectrum remains unchanged; here the infrared-color film shows stressed leaves as a lighter red to white.

Significantly, the loss of plant vigor may be detected on color-infrared photography several days before it becomes visible to observers in the field. Plants in the coniferous group, on the other hand, display somewhat different characteristics. Most importantly the loss of infrared reflectance usually does not show up before visible signs of death occur. Dying needles of conifers, which appear yellow-green on color transparencies, assume a pink hue on Ektachrome Infrared transparencies; yellow trees appear whitish; and yellow-red and red trees show up yellow on the infrared-color film [6]. Nevertheless discoloration of the crown of a tree is far more easily detected by aerial photographic methods including color infrared than by observers on the ground.

The reason why color-infrared film responds to even the slightest change in plant vigor is only now becoming understood. It is not, for example, simply a function of the chlorophyll-bearing tissue. Lichens, which are living organisms of simple morphology containing chlorophyll, do not photograph red, but blue-green to whitish. The critical element appears to be the cell-wall/air space interface within the mesophyll tissue. In a healthy mature leaf the mesophyll tissue tends to be similar to a sponge, that is, full air spaces. In this condition the tissue is an efficient reflector of radiant energy, in particular the near-infrared wavelengths [7]. If for any reason this condition is disturbed, such as a reduction in the water content, the plant may begin to lose vigor. This will cause the mesophyll to collapse, impairing its ability to reflect the near-infrared wavelengths. The change in the infrared reflectivity is immediate and may occur a day or two before a corresponding change occurs in the chlorophyll. Leaves, therefore, which have little or no mesophyll (conifers) reflect much less infrared.

Although color infrared's ability to detect changes in plant vigor has received most of the attention, the film has other qualities which made it useful for land use planners. One of these is its superior haze penetration capability. In accordance with Rayleigh's Law the longer wavelengths of the infrared portion of the spectrum are scattered less by haze particles than the shorter wavelengths. And since the use of a yellow filter eliminates the short blue wavelengths anyway, the film can consistently provide sharper photos of hazy metropolitan areas than can conventional color films.

Another factor which contributes to the effectiveness of color-infrared photography is the contrast enhancement of scene elements. This is 'brought about by the higher (and lower) reflectances of certain objects in the infrared spectral region than in the pan film sensitivity range.' [8]. Contrast is perhaps greatest between cultural and natural features, and thus the film is unusually valuable for land use mapping, particularly in suburban and rural areas.

Still, there are limitations in the usefulness of color-infrared film which must be kept in mind. The fact that some kind of filter must be used with color-infrared film is one such example. In fair dry weather a yellow filter is employed to absorb the blue and violet wavelengths to which all three emulsion layers are sensitive. But, if there is excessive haze or water vapor in the atmosphere even heavier filtration may be necessary. The use of these filters necessitates the requirement of sufficient light to sensitize the film. Unfortunately, at certain periods of the year and at certain latitudes these conditions may be found only infrequently.

Furthermore, caution must be used because of the damaging effect age has upon color-infrared film. The infrared-sensitive cyan layer tends to decrease in speed as the film ages while the speed of the green-sensitive yellow layer tends to increase. As the film ages its color balance shifts toward cyan. If the film is kept at room temperature its balance will eventually pass the point of optimum discrimination. These effects can be reduced considerably by refrigeration and eliminated almost entirely by storage in a freezer [9].

Finally, there is the cost factor. Color-infrared film is several times more expensive to purchase than pan films and is also more expensive to process (Table 3). Still, no film serves all purposes and each has its advantages and disadvantages. There will be occasions, then, when the type and quality of data acquired by color-infrared film will outweigh whatever the disadvantages in terms of handling and cost. And to repeat, the costs of film and processing may represent an almost negligible fraction of the total costs of an overflight.

2.5 Photographic Products

Regardless of the type of film employed, the final product may be delivered either as a positive paper print or a positive transparency. In terms of cost and durability, paper prints are superior to transparencies; most black-and-white photos are produced in this format. Paper prints can be produced as contact prints, that is, produced at the same scale as the negatives, or enlarged to a more convenient working scale. They can be readily taken into the field, and if in so doing are damaged, can be replaced inexpensively assuming the negatives from which the originals were produced are still in existence. Paper prints may be produced in either a matte, semi-matte, or glossy finish. A matte finish is dull and is good for annotating on the photos; a glossy finish is poor for purposes of annotation but best for illustrations. The semi-matte finish is a compromise and is the prefered finish for most general uses.

Table 3. Costs of aerial films*

Film type	Size (ins.)	Price range ($)	Approximate processing costs ($)
Black-and-white	$9\frac{1}{2} \times 250$ ft.	260–285	0.40 per ft.
Black-and-white IR	$9\frac{1}{2} \times 250$ ft.	520–540	0.40 per ft.
Color negative	$9\frac{1}{2} \times 200$ ft.	740–760	1.00 per ft.
Color reversal	$9\frac{1}{2} \times 200$ ft.	700–715	1.00 per ft.
Color IR	$9\frac{1}{2} \times 200$ ft.	825–840	1.00 per ft.

*Prices as of May, 1984.

Transparencies, though more expensive to produce and requiring greater care in their handling, are definitely superior to paper prints both in terms of ease and accuracy of interpretation. This is a result of the generally higher resolution qualities of transparencies as compared with paper prints combined with the fact that transparencies must be viewed over a light table. Illumination of photos from below has been amply demonstrated as being a more effective method of interpreting photos than illuminating them from above. All films may be produced as transparencies either from negatives or, as in the case of reversal films such as Ektachrome Infrared, directly from the original film exposed in the camera. It should be emphasized that when Ektachrome Infrared transparencies are being used, great care should be taken in their handling. Unless the film has been duplicated, there is no way of replacing it if it is damaged. Reversal films have no negatives. One way to protect such films, if heavy use is expected, is to cut up the roll into individual frames and place each frame in a plastic sleeve. This format also provides for easy storage.

3. The geometry of aerial photos

3.1 Introduction

Aerial photos are usually classified according to the inclination of the camera axis at the moment of exposure. Thus, photos are termed either verticals, low obliques, or high obliques. Verticals are taken with the optical axis of the camera in a vertical position (Fig. 11); more commonly, however, the axis is 1–2° from vertical because of uncompensated motions in the aircraft. Oblique photos, in contrast to verticals, are taken with the optical axis of the camera inclined. Those taken with the optical axis inclined enough that the horizon is visible on the photos are referred to as high obliques, those on which the horizon is not visible are referred to as low obliques. While each type of photo has its uses, the utility of vertical photos far exceeds that of obliques.

3.1.2. *Verticals*. Vertical photos are preferred by land planners because of the ease with which information derived from the photos can be transferred to maps and other data bases. Verticals record information from a rectangular-shaped portion of the earth's surface at a scale that is generally constant throughout the photo. While verticals may be acquired singly or in pairs they are usually taken in sequence along a flightline. Photos on the same flightline ordinarily overlap one another by about 60 percent while those on adjacent flightlines overlap by 30 percent. The purpose of this overlap, termed endlap in the former instance, sidelap in the latter, is to provide complete coverage of an area and make possible stereoscopic or"three-dimensional' analysis, a technique to be discussed in the following chapter.

3.1.3. *Obliques*. While it is possible to employ oblique photos for mapping purposes and even to examine them stereoscopically, their use is complicated by the problem of scale. Unlike verticals which have a constant scale throughout, oblique photos have a constant scale only along a line parallel to the flightpath. At right angles to the line of flight the scale changes continuously according to the laws of perspective. The scale is greatest in the foreground and decreases in the direction of the horizon.

Scale is not the only problem, however. On high obliques in particular, haze

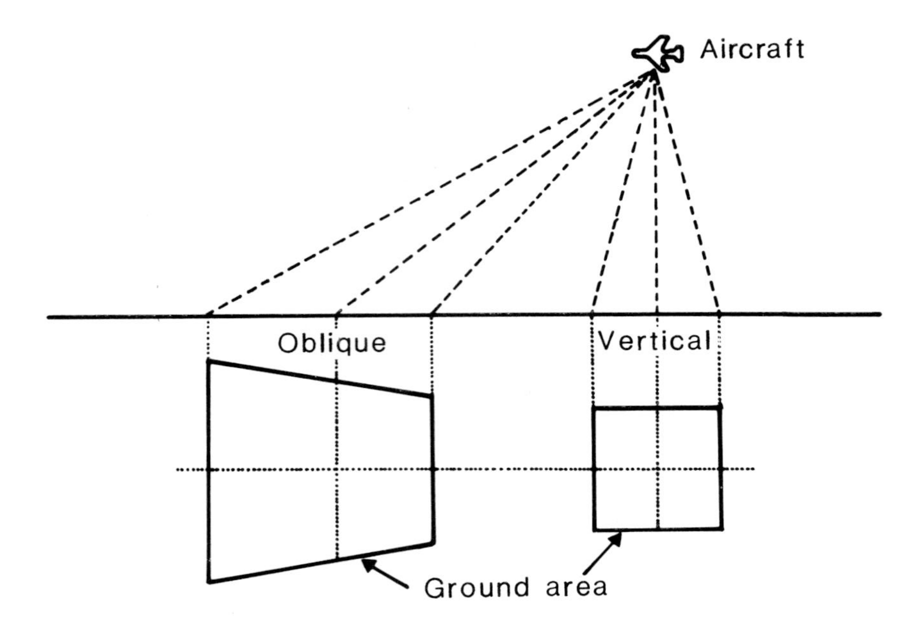

Fig. 11. Comparison of ground coverage for vertical and oblique photos.

may present a further complication. Because of the extreme thickness of the atmosphere through which the reflected radiation has to pass, details of the land surface may become completely lost near the horizon.

Oblique photos cover a larger ground area than the typical vertical and they present the terrain in a more familiar perspective. Furthermore, it is not always necessary to employ expensive camera systems to acquire obliques. The growing literature on the use of 35 mm cameras attests to this fact. While oblique photos may not be required during the analysis stage of a project, they may become extremely valuable in later stages as illustrations for final reports and public presentations.

3.2 Scale and Resolution

The amount of information that can potentially be extracted from an aerial photo is in large part a function of the scale and the resolution of the photo. Scale is the ratio of a unit distance on the photo image to the actual distance on the ground. Thus, a scale of 1 : 20,000 means that one inch or centimeter on the image is the equivalent of 20,000 inches or centimeters on the ground. The simplest way to determine the scale of a set of photos is to measure the distance between two features on one of the photos and measure the corresponding distance on the ground. The scale can be computed as a ratio of the photo distance to the ground distance, or

$$RF = \frac{\text{Photo distance}}{\text{Ground distance}}$$

thus, for photos on which a football field measures 0.03 feet, the scale or representative fraction would be

$$\frac{0.03'}{300} \quad \text{or } 1:10{,}000.$$

On vertical aerial photos scale is expressed as a ratio between the focal length of the camera and the flying height of the aircraft above the terrain, or

$$RF = \frac{f}{H\text{-}h} \quad \text{where} \quad \begin{aligned} f &= \text{focal length} \\ H &= \text{height of camera above sea level} \\ h &= \text{average terrain elevation} \end{aligned}$$

Figure 12 illustrates these relationships. Focal length is represented by the dashed line Lo where L represents the camera lens and the line boa the film negative. Scale is inversely proportional to focal length. The height of an aircraft (or camera lens) above the terrain is represented by the distance OL and is determined by subtracting the average elevation of the terrain from the altitude of the aircraft above sea level. Scale is directly proportional to flying height. An aircraft flying at 12,000 feet above sea level over terrain averaging 2,000 feet above sea level would acquire vertical photos of a scale 1:10,000 if a 12-inch focal length camera were employed, 1:20,000 if the focal length were 6 inches.

Like scale, the ground area covered by a vertical photo is a function of focal length and altitude. Aerial coverage is also inversely proportional to focal length and directly proportional to flying height. There is, however, a trade-off between scale and aerial coverage. The larger the scale of a set of photos (and the more detail that can be extracted from them) the larger number it requires to provide complete ground coverage of a given area. This may mean an added expense in acquiring the photos, not to mention the problem of manipulating and interpreting the larger number of photos.

The scale required for a set of photos depends ultimately upon the specific task at hand. For certain engineering purposes scales of 1:5,000 and even larger may be needed. For most general interpretation purposes scales of 1:15,000 to 1:20,000 appear optimum, particularly if the photos are to be black-and-white paper prints. Scales smaller than 1:20,000 are generally recommended only if the area to be examined is fairly large. In this instance it is recommended that the photos be in the form of transparencies because of their higher resolution. Table 4 provides a list of the more commonly employed scales and the ground area covered by each.

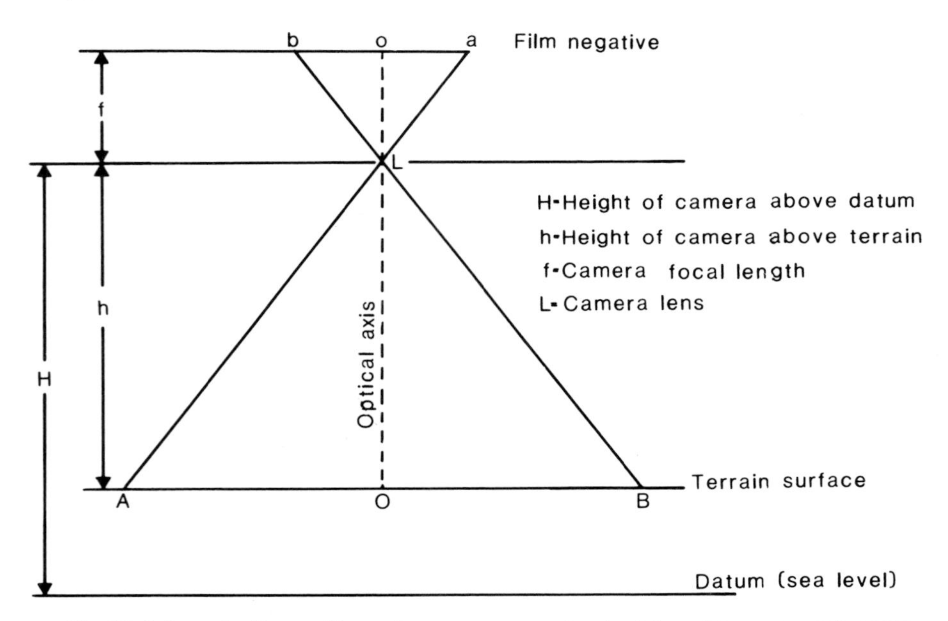

Fig. 12. Schematic diagram illustrating components of scale. Adapted from Westfall (1973).

Table 4. Altitude, width of ground strip, and aerial coverage by selected scales

Photo scale	Flying height 6″ (152 mm) focal length	Width of ground strip in miles (km)	Area covered on 9″ print in sq. mis. (km²)
1:5000	2500′ (760 m)	0.71 (1.14)	0.50 (1.30)
1:10,000	5000′ (1520 m)	1.42 (2.29)	2.01 (5.23)
1:20,000	10,000′ (3040 m)	2.84 (4.57)	8.07 (20.90)
1:40,000	20,000′ (6080 m)	5.68 (9.14)	32.28 (83.61)
1:63,360	31,680′ (9629 m)	9.00 (14.40)	81.00 (209.79)
1:80,000	40,000′ (12,160 m)	11.43 (18.29)	130.64 (334.45)
1:100,000	50,000′ (15,200 m)	14.28 (22.86)	203.91 (522.58)

Finally, it should be pointed out that while a set of photos may be said to have a particular scale, for example 1:20,000, the figure used is only an approximation. The actual scale of the individual photos may vary greatly. Some of the variation may result from the inability of the pilot to maintain the aircraft at a constant altitude because of atmospheric turbulence. More commonly, scale variations are due to differences in relief. The 1:20,000 scale, mentioned as an example, refers to the scale of the datum plane (A-B in Fig. 12). Areas of high relief have the effect of altering the orientation and distances between terrain features. These variations in scale can be particularly troublesome if the photos are to be matched up or if measurements of some kind are to be made from them.

Resolution refers to the optical quality of an aerial photo; it is determined primarily by film speed, or rather the grain size of the halide crystals in the film emulsion. The resolution or resolving power of an aerial film is defined by the width of the narrowest pair of lines that can be resolved separately on a black-and-white test target. Tests are most commonly conducted in the laboratory but on occasion are conducted in the field. The results are expressed as the maximum number of lines that can be resolved per millimeter. Resolving power is limited by grain size, that is, the larger the halide crystals the larger the minimum size of the terrain features recorded on the negative. Grain sizes vary from about 0.1 to 5.0 m with the faster films, which are those more sensitive to light, having the larger grain sizes. Thus, there is a trade-off between film speed and resolution. Aerial films have grain sizes of about 1.0 m because of their need for high resolution [10].

The resolving power of a film is also a function of the contrast in scene elements, the greater the contrast the higher the resolution. Test targets are usually black or white and various shades of gray. Since contrast is so important the resolving power of a film is specified at some contrast ratio such as 150 lines/mm at 6.3 : 1. The latter figure is a contrast ratio between the lines and the background. Common contrast ratios are 1.6 : 1, 6.3 : 1, and 100 : 1. Low contrast ratios (1.6 : 1) most realistically represent actual terrain contrasts.

The resolution of any aerial photo is always less than the resolving power of the film, particularly if the resolving power has been determined by laboratory tests. Resolution is influenced additionally by such considerations as the quality and characteristics of the camera optics, the motions of the aircraft, and the condition of the atmosphere (clouds, haze, etc.). And finally, the photos being used may also be several generations removed from the original negatives and with each generation some resolution is lost.

As a postscript to the discussion it should be stated that resolution, while described above in very technical terms, is not always used in such a precise manner. Many photointerpreters define resolution more in terms of ground resolution and express it simply as a function of the smallest object clearly resolved on the photo, i.e., three-foot resolution.

3.3 Image Displacement

As orthogonal projections, maps depict features in their true horizontal relationships to one another. This is not the case with aerial photos. As perspective projections aerial photos suffer from a number of displacements that can make difficult, if not impossible, the accurate measurement of distances or the precise transfer of land use data to maps. These displacements may be grouped into three categories according to their origin and importance: those related to the camera system, those related to tilt in the optical axis of the camera, and those related to relief. To further understand these displacements three terms must be introduced which refer to three different photo centers: the principal point, the nadir, and the isocenter (Fig. 13). The principal point is formed where a perpendicular drawn through the center of the lens intersects the negative plane; lens distortion increases radially from the principal point. The nadir is the point directly beneath the camera lens at the time of exposure; relief displacement increases radially from the nadir. The isocenter is the point midway between the principal point and the nadir; tilt displacement radiates from the isocenter. On a truly vertical photo, these three points would be identical.

3.3.2. *Camera Distortions*. Distortions originating with the camera system have a variety of causes – poor quality optics, a faulty shutter on the camera, or damaged film. The most common is the distortion caused by the lens system. In all cameras incident light passing through the lens system is deflected to some degree from its geometrically correct path. This, as well as the other distortions caused by the camera system, have largely been eliminated through the use of precision mapping cameras.

3.3.3. *Tilt*. Displacement caused by tilt in the optical axis of the camera is found in most aerial photos because of the difficulty in maintaining an aircraft on a perfectly horizontal flightpath. In Figure 13 the optical axis of the camera, POp, is represented as being tilted. Where this line intersects the negative plane, p, the point is referred to as the principal point and is located at the exact center of the negative; where the line intersects the ground, P, the point is known as the ground principal point. The point directly beneath the camera, the nadir, is represented on the ground as N and on the negative as n. While the lines POp and NOn would be identical on a perfectly vertical photo, they are not so on a tilted photo. The principal point, p, is still the center of the tilted photo but it is removed from the nadir, n, by a distance determined by the optical angle pOn. The focus of the tilt in this figure is located at the isocenter, i, and is defined as the point where a line bisecting angle pOn intersects the negative plane. Displacement occurs radially from the isocenter. On the upper side of the tilt plane images are displaced toward the isocenter, on the lower side of the tilt plane away from the isocenter. No displacement occurs along the axis of tilt.

The effect of tilt displacement on aerial photos may best be visualized by reference to Figure 14. To simplify the illustration tilt is presented as acting only along the direction of flight. In reality tilt is random and far more complicated.

As long as the amount of photographic tilt remains less than 2 or 3° it can probably be tolerated for most purposes; it may even pass undetected. Tilt should not be allowed to exceed 3°. Photos taken with the optical angle greater than 3° are technically low obliques and must be handled as such. Agencies or individuals contracting with an aerial survey company to acquire photos of a given area should make one of the conditions of the contract that tilt not exceed

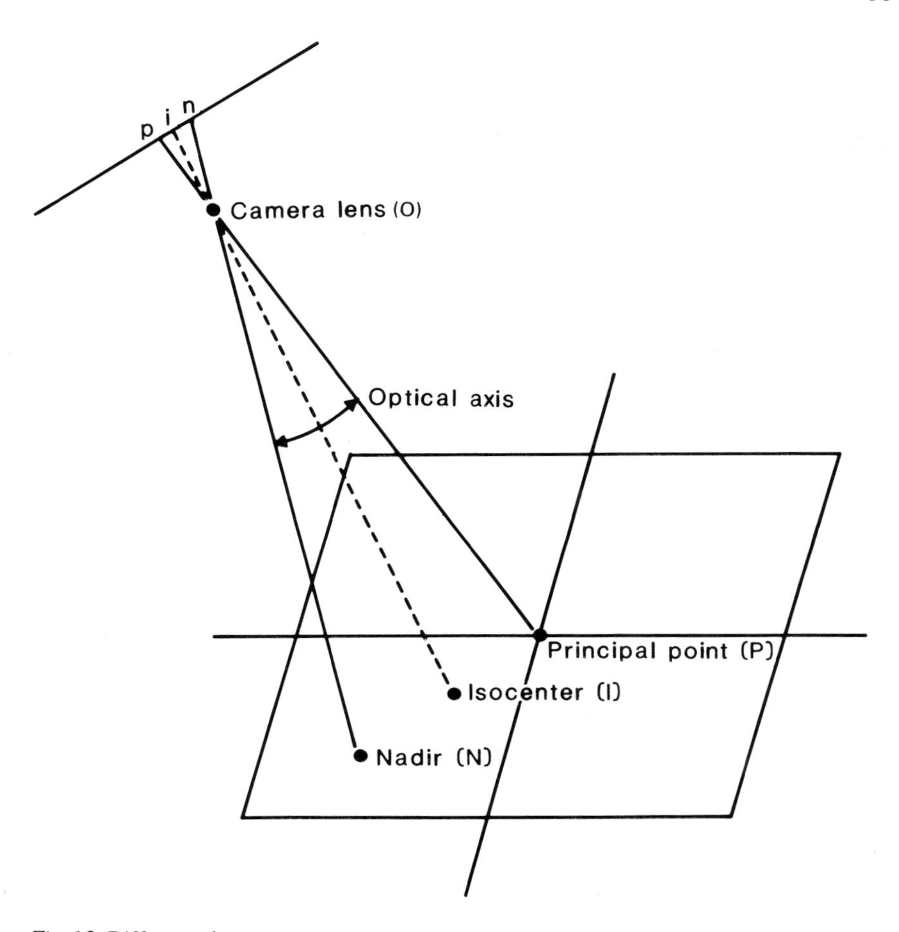

Fig. 13. Different photo centers.

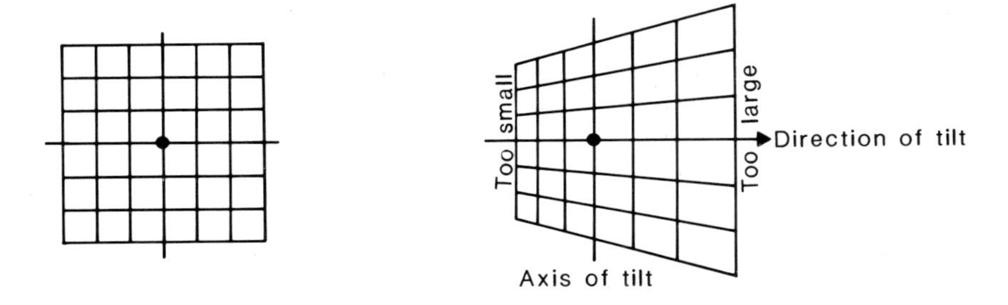

Fig. 14. Effect of tilt displacement.

3° for any individual photo nor an average of 1–2° for the mission as a whole.

3.3.4. *Relief.* The most significant source of distortion found in aerial photos is caused by differences in relief and is referred to as relief or topographic displacement. Relief displacement can best be described as 'the apparent movement of an object's image on the photograph in relation to its true position on the ground' [11]. This movement results from the perspective view seen by the camera when the optical axis is in a vertical position. Relief features or tall buildings rising above the datum plane are displaced outward from the nadir while those features extending below the datum plane are displaced inward. The amount of displacement increases radially from the nadir. There is no displacement at the nadir. In Figure 15 hilltop A′, which on a planimetric map would be located at A, is displaced outward from the nadir (identical to principal point p) on the negative by the distance a–a′. Valley B′, on the other hand, which extends below the datum plane and on a map would appear at B, is displaced inward on the negative by the distance b–b′. Point C, located directly on the datum plane, appears on the negative (c) in its proper location relative to the principal point.

Relief displacement is a function of distance from nadir to the top of the displaced object, height of displaced object, and flying height of the aircraft above the terrain. It can be computed as follows:

$$\text{Relief displacement} = r\frac{h}{H}$$

where
r = radial distance from nadir to top of displaced object
h = height of displaced object
H = altitude of aircraft above terrain

Relief displacement is inversely proportional to flying height (there is little relief displacement on imagery acquired from orbiting satellites) and directly proportional to object height and distance from the nadir.

Although relief displacement is considered a type of distortion because of the difficulties it creates for measuring horizontal distances, it also makes possible stereoscopic viewing and the estimation of heights of displaced objects. To estimate heights several conditions must be met. First, the height of the aircraft above the base of the object must be known precisely. Second, both the top and the base of the object must be clearly visible. And third, the distance from the nadir to the top of the displaced object must be great enough to be measurable. If these conditions are met, the height of an object can be calculated using the following equation:

$$h = \frac{d}{r}(H\text{-}h)$$

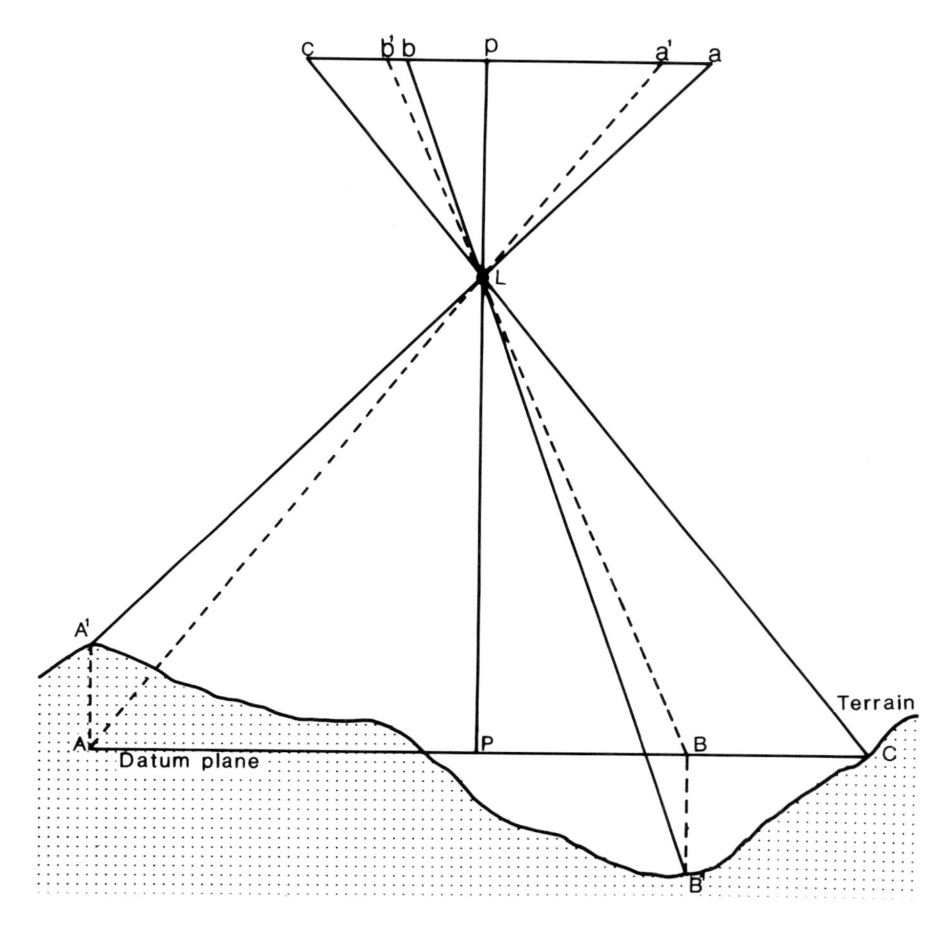

Fig. 15. Relief displacement on true aerial photo. Adapted from Westfall (1973).

where d = length of displaced object
r = radial distance from nadir to top of displaced object
H = altitude of aircraft above sea level
h = height of object base above sea level

The relief displacements described are those found on vertical aerial photos. When photos suffer from tilt as well as relief displacement the resulting geometry is very complex and makes difficult the use of such photos for mapping purposes requiring high accuracy. Photos to be used for mensuration purposes must be rectified or corrected for various distortions.

3.4 Rectification of Aerial Photos

The technical term generally given to the process of plotting planimetrically correct information from aerial photos is rectification, although as Kilford has observed, this is an unfortunate choice of term 'because photographic rectification is a method of producing a print free from tilt and having no distortions due to the lens' [12]. By this latter definition rectification can be accomplished by a variety of methods, several of which are relatively simple. In its more general meaning, that is, in the sense of correcting for both tilt and relief, rectification can be carried out only through the use of expensive plotting machines.

Small amounts of tilt may go undetected or, at least, pose no problems for minor map revisions. However, the effects of tilt and minor scale variations may be largely eliminated by the use of one of several devices designed especially for the transfer of data from an unrectified photo to a map base. The vertical sketchmaster, as one example, employs a single vertical photo which is fastened to the photocarrier (Fig. 16). The device is placed over the map base. Tilt distortions are corrected by tilting the photocarrier, while scale is made to correspond to the map by raising the photocarrier up and down; screw settings on the legs allow for these corrections. A semi-reflecting mirror makes it possible for the user to view the photo superimposed on the map. Revisions to the map can be accomplished simply by tracing the superimposed features.

A more sophisticated, as well as more expensive, instrument is Bausch and Lomb's Zoom Transfer Scope (ZTS). The ZTS allows the user to view two images simultaneously – one superimposed over the other. Most commonly, the ZTS is used to superimpose an aerial photo onto a map of the same area for the purpose of transferring information from the photo to the map. In the process of accomplishing this, the ZTS can correct for scale differences as well as certain optical distortions. For example, if the scale of the photo is smaller than that of the map, the zoom magnifier may be increased from 1 × up to as high as 14 × or to the appropriate map scale. If, on the other hand, the photo scale is larger than the map, the standard 1 × lens can be replaced by either a 2 × or 4 × in order to match the two scales. In addition to the zoom magnifier and different lens sets, an anamorphic control on the ZTS allows for the correction of geometric anomalies resulting from tilt, lens distortion, or film shrinkage. The control enlarges the image in any one direction, the effect of which is to make corrections without complicated rectification equipment. Another application of the ZTS is to superimpose two photos of the same area but taken on different dates over one another. This allows for a rapid determination of land use change between the two photo dates. The ZTS can accommodate both paper prints and transparencies. The newest models such as the one pictured in Figure 17 allow for stereoscopic (three-dimensional) viewing as well as monoscopic viewing.

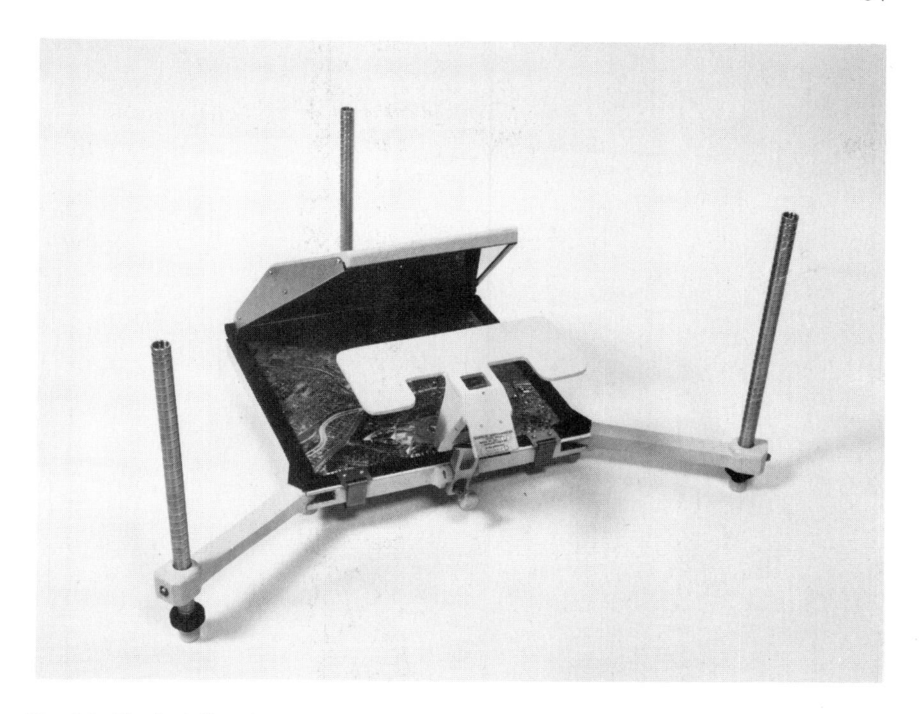

Fig. 16. Vertical Sketchmaster, Type 260GE. Courtesy Gordon enterprises.

Both of these devices are capable of correcting for tilt and scale differences when data are being transferred to a map base. It is also possible to produce positive prints free of tilt and lens distortions. Rectified prints are made with an optical rectifier. The typical rectifier consists of a projection lamp, lens, and movable map plane. Light from the lamp is projected through a film negative, the rays passing through the lens to be focused on the map plane holding print paper. The lens, similar to that on an aerial camera, acts to eliminate distortions as the negative image is projected back through the lens. Tilt is removed by tilting the map plane holding the print paper. Thus, the image projected through the lens exposes the print paper on the map plane tilted to compensate for the tilt in the original negative. The positive print produced is then free of lens and tilt errors but not relief displacement. Nevertheless, if the terrain covered by the set of rectified photos is relatively flat, the photos will contain few distortions of any kind. Thus, a resonably accurate map can be produced simply by tracing the various terrain features onto an overlay to the photos.

If new maps or map revisions are to be made of areas with considerable relief, the techniques just described will be inadequate. What is needed is a device that will simultaneously correct for both tilt and relief displacements. Such a device is called a stereoplotter. There are a number of different types on the market.

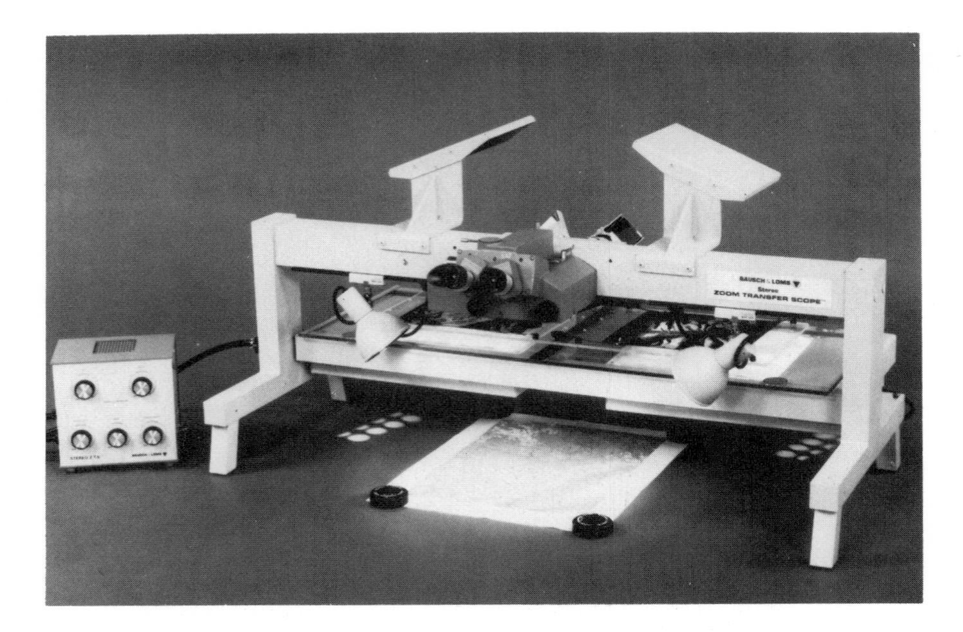

Fig. 17. Stereo Zoom Transfer Scope. Courtesy Bausch and Lomb.

Most types will produce a highly accurate and detailed plot or map with contour lines if desired.

Stereoplotters are sophisticated and expensive pieces of equipment, and it is not the intent of this author, nor is it even necessary, to describe the stereoplotting process in any but the most cursory manner. The best advice, if one is contemplating the need for an accurate planimetric map, is to consult with a representative of a reputable aerial survey company with considerable experience in the production of maps from stereoplotters.

A stereoplotter is a machine that creates a stereomodel from which terrain measurements can be made and projected orthographically to a map sheet. The map produced in this manner may contain both contour lines and planimetric detail. In accomplishing this, the stereoplotter corrects for errors resulting from lens distortion, tilt, and relief. The plotter thus transforms photo coordinates into correct map coordinates.

The stereomodel is created by projecting light through two overlapping diapositives (glass-based positives) onto a common plane for viewing. In projecting these images onto the viewing plane the process by which they were originally acquired is reversed, that is, the images formed by focusing reflected light from the earth's surface through the camera lens onto the negative plane are now themselves projected back through a lens, similar to that on an aerial camera, onto the viewing plane where the overlapping images produce the stereomodel. By this process some of the error caused by tilt and lens distortion

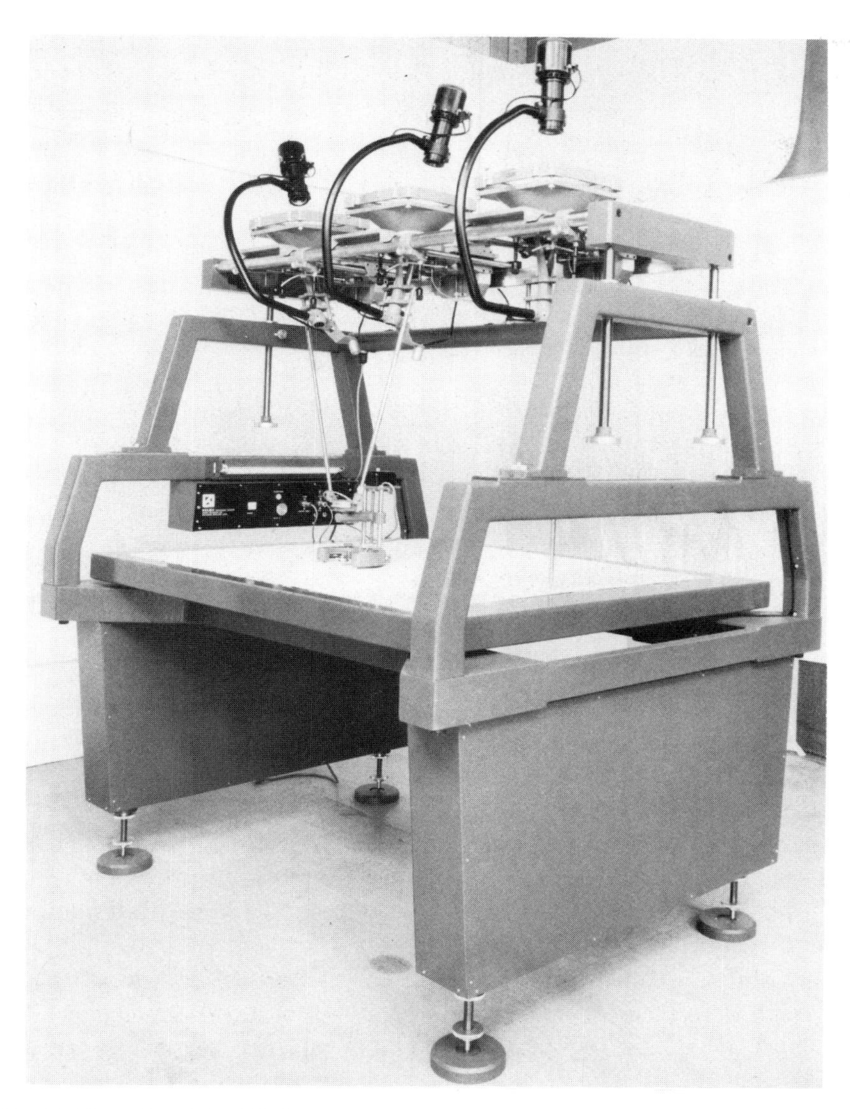

Fig. 18. Kelsh Model KPP-3B Stereo Plotter System. Courtesy Danko Arlington, Inc.

is removed. The model is then corrected for scale and orientated with respect to ground datum. This is done by tying the model into a number of ground control points whose exact coordinates and elevations are known.

Measurements made from the stereomodel are carried out by means of a floating dot which is actually a point of light at the center of the tracing table platen. The height of the floating dot can be calibrated to the height of the terrain. When the stereomodel is viewed through the plotter's optics the floating dot is visible against the terrain and, if the platen is moved around the model properly, keeping the floating dot in contact with the terrain surface, a predetermined contour line will be traced out. At the same time the contour line is being traced out on the model, a plotting arm is drawing the contour line on the map. Similarly, other features on the photos can be correctly transferred to the map. Increasingly today, however, stereoplotters are becoming completely computerized. Not only can all the corrections be made automatically on such systems but more importantly the contour and planimetric data can be digitized and stored on tape. Computer maps displayed some or all of these data can be generated at any time.

A number of photographic products are available that have had most, if not all, distortions removed. The most important of these are controlled mosaics and orthophotos.

3.4.2. *Controlled Mosaic*. A controlled mosaic is an assemblage of photos that have been cut and fastened together in such a way as to produce a single large photo meeting certain planimetric standards. Such a mosaic is made from good quality photos which have been rectified to eliminate tile distortions. The photos are then cut and fitted together; only the centers of the photos are used to eliminate as much relief displacement as possible. The pieces are next glued to a solid backing during which process a number of locations on the mosaic are tied into ground control points whose exact locations have been predetermined with great accuracy. The number of ground control points used in the preparation of the mosaic determines its accuracy and whether it is actually a controlled or semi-controlled mosaic. Since the preparation of a controlled mosaic is expensive it is advisable not to have one prepared unless completely necessary. If such a mosaic is deemed necessary it is equally advisable to have it prepared by a reputable aerial survey or engineering firm.

There are other types of mosaics in the preparation of which little or no attempt is made to correct for distortions. These are collectively referred to as uncontrolled mosaics. In the simplest of these, the index mosaic, individual photos are laid out by flightline and stapled, or fastened in some way, to a piece of composition board. The serial numbers of each photo are visible so one may determine, and order if necessary, the specific photos one is interested in. A second type of uncontrolled mosaic is one employing every other photo in order to create a regional base map. The photos are stapled to a sturdy piece of composition board, and the analysis is done on an overlay to this composite map. Where necessary stereo analysis can be conducted using the alternate photos not required in the preparation of the mosaic. And finally, it is possible to create an uncontrolled mosaic having the appearance of a single, large photo. This is done by cutting the individual photos, sanding the edges, fitting them together, and gluing them to a solid base. To reduce some of the distortions only the centers of photos are used. While uncontrolled mosaics, they are often sufficiently accurate for many land planning purposes and are not very costly to

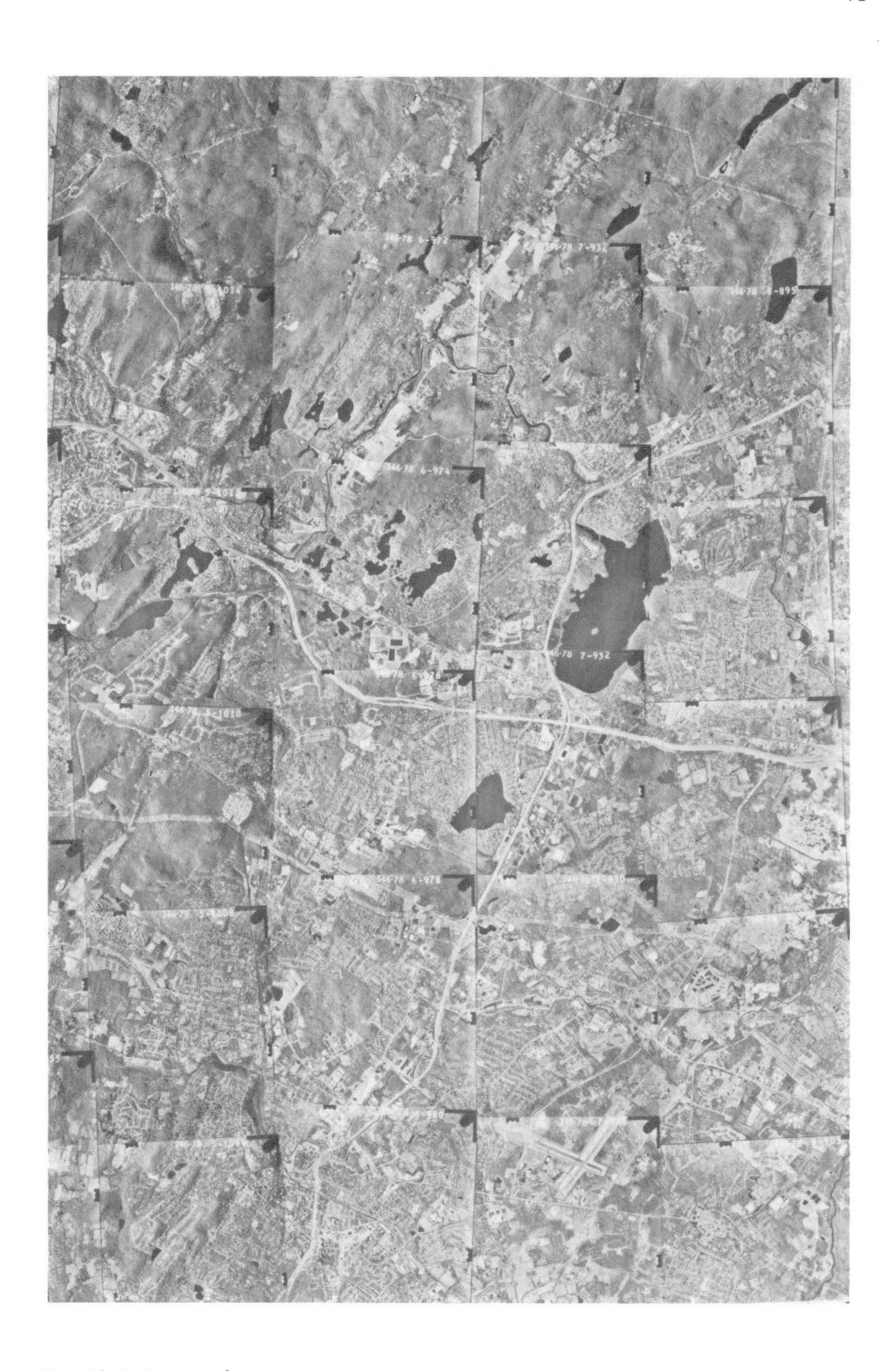

Fig. 19. Index mosaic.

produce. They can even be overprinted with annotations and rephotographed producing a relatively inexpensive photomap.

3.4.3. *Orthophotographs*. Similar to the controlled mosaic but embodying even greater accuracies is the orthophotograph or orthophoto. The term 'orthophoto' refers to aerial photos from which variations in scale as well as displacements resulting from tilt and relief have been removed. The result is an aerial photo with its wealth of detail having the same planimetric accuracy as a topographic line map.

Orthophotos are produced by optical-processing machines, called orthophotoscopes, which are connected to stereoplotters. Employing a process known as differential rectification, scale variations and relief displacement are removed from the original negatives on a point-by-point basis as the orthophotoscope scans the stereomodel. The negatives corrected in this manner are fitted together to produce the final photo of the desired area. As one might expect, orthophotos, like controlled mosaics, are expensive to produce though they are becoming less so.

Increasingly orthophotos are being used by states and local communities for tax mapping purposes. They are preferred to line maps because of the ease with which property lines can be related to terrain features such as streams, stone walls, and treelines. Distances and areas can also be calculated from orthophotos thus reducing the errors involved in the transfer of data from field notes, deeds, and other maps.

There are limitations to orthophotos. In general, the quality of orthophotos is lower than that of standard photography. There is the problem of keeping the photos clean and free of scratches during the reproduction process. There are also problems of image blurring, visible scanlines, and matching tones between scanning strips. Finally, there is the loss of resolution between the orthophoto and the original photography. The orthophoto is usually several generations removed from the original [13].

Orthophotos do not necessarily contain topographical information although orthophotoscopes are now capable of generating contour data automatically. The U.S. Geological Survey is presently publishing a series of orthophotomaps or quads which have the same planimetric accuracies as the standard topographic maps.

4. Fundamentals of photointerpretation

4.1 Introduction

The effective use of aerial photos requires an ability to do two things – first, to detect and identify the individual features visible on the photo, and second, to draw conclusions on the basis of what is observed. To the first, the term photo-reading has been assigned, to the second, photo-interpretation. While some may take exception to this distinction, it may be a helpful one to those new in the use of aerial photos. With time the distinction will admittedly blur as recognition and interpretation become almost simultaneous processes.

The recognition of features is a continually occurring process in our daily lives although not always a conscious one. It becomes a conscious process when we look at an aerial photo for the first time because the vertical perspective makes even those features most familiar to us on the ground difficult to identify. Further contributing to the problem is photo scale; the small size of features as they appear on a photo obscures many details critical to their identification. And, finally, on black-and-white photos there is also an absence of color. The color of an object can be an extremely important aid in its identification. In spite of these obstacles the process of reading photos can be readily learned; it simply requires practice working with photos. Photo-reading is primarily a matter of accustoming oneself to viewing familiar features from an unfamiliar perspective. To do so requires the use of attributes different from those generally employed in the identification or recognition of features on the ground. These attributes, of which seven are commonly recognized, can be described as follows.

4.1.2. *Size.* This attribute is more important in the relative sense than in the absolute sense although occasions may arise when knowing the actual dimensions of any object may hasten its identification. More often the problem is one of confusing a dwelling with an equipment shed, a possibility enhanced by the not infrequent use of several sets of photography having different scales. In such instances it is advisable to view a number of known objects on each of the photo sets to ascertain the relative size of all the landscape features.

4.1.3. *Shape.* Shapes of the most familiar features on the ground may appear

Fig. 20. Interstate 89 and portion of Lebanon Regional Airport. Lebanon, New Hampshire, November 1968. Scale 1 : 6000. Courtesy Photographic Interpretation Corporation.

startlingly different from the air. On the other hand, many features have such distinctive shapes, e.g., a soccer field, that this attribute alone is sufficient for identification. As a general rule, man-made features have very regular shapes while natural features have irregular shapes. In Figure 20, for example, compare the meandering path of the stream with the relative preciseness of the airport runway, the interstate highway, and the powertrace running through the woods.

4.1.4. *Shadow.* On aerial photographs shadows usually represent an absence of information. To reduce their effect, photos are acquired either just before or just after midday. There are instances, however, when shadows can be invaluable to the identification process. In Figure 21 notice the shadows of the trees located along the road and along the edges of the fields. These shadows may provide an important means of identifying the species of trees found in this area.

4.1.5. *Tone.* On black-and-white photos tone or greyness is a function of sun angle, inclination of the surface, and the condition of the atmosphere. Thus, the movement of the aircraft itself may cause an object to appear nearly black on

Fig. 21. Wooded area of Etna, New Hampshire. Scale 1:6000. Courtesy Photographic Interpretation Corporation.

one photo and nearly white on the next. In general, features cannot be associated with a specific grey tone. Relative differences are important in discriminating between certain features – unpaved versus paved roads, coniferous versus deciduous trees, moist versus dry soils. The range of grey tones in Figure 22 is a reflection of vegetation and terrain differences.

4.1.6. *Pattern.* Pattern refers to the natural or man-made arrangement of individually visible features on the landscape. As with shape, natural patterns tend to be less regular than man-made patterns. The importance of pattern is what it may reveal about an area. For example, a dendritic drainage pattern formed by a stream and its tributaries may suggest the area is one of sedimentary rocks; an orchard with its individual trees laid out in a particular pattern reveals something of the farming activity taking place. Can you provide an explanation for the 'frame' around the field pictured in Figure 23, or the banded pattern to the woods in Figure 24?

4.1.7. *Texture.* In contrast to pattern where the individual objects are visible,

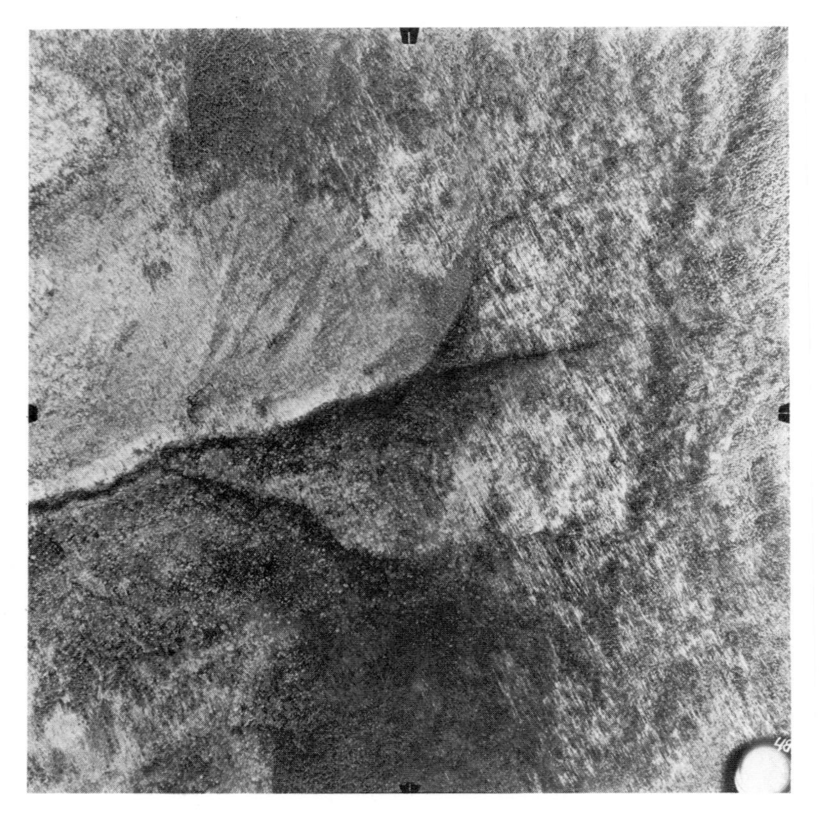

Fig. 22. Heavily wooded New Hampshire landscape. Scale 1:6000. Courtesy Photographic Interpretation Corporation.

texture refers to features comprised of objects too small to be seen. The effect is a surface that to the observer appears either smooth or rough. Again the term is a relative one since texture varies with scale. Nevertheless, sand is generally considered a finely textured surface, pastureland moderately textured, and woodland coarse textured. A variety of textured surfaces are illustrated in Figure 24.

4.1.8. *Relative location.* Location can be an important identification factor particularly in urbanized areas where many activities require locations meeting certain prerequisites. A large, one-storey building adjacent to a railroad track is more apt to be industrial than commercial. Commercial activities require easy access by large numbers of shoppers, so they are usually located along busy streets or highways. Within residential areas small commercial establishments such as variety stores, pharmacies, and dry-cleaning establishments may prefer locations at the intersections of main thoroughfares for the same reasons.

Photointerpretation is the process that begins once one has mastered the

Fig. 23. Alternating fields and woodlands, typical of northern New England. Scale 1:6000. Courtesy Photographic Interpretation Corporation.

technique of photo-reading. It implies an ability to assess the significance of what is being observed and to draw conclusions from it. Thus, photointerpretation requires specialized training in some discipline – engineering, geography, geology, or planning. For this reason it is far more efficient to train a specialist in the use of aerial photos than it is to train a non-specialist. Aerial photo-interpretation is a tool, and as with any tool, its efficiency is proportional to the experience of the person using it.

4.2 Stereoscopic Analysis

Aerial photographs can be examined in two ways – monoscopically and stereoscopically. Monoscopic examination involves the viewing of a single photo. Such

Fig. 24. New Hampshire country road. Scale 1:6000. Courtesy Photographic Interpretation Corporation.

a viewing technique, however, utilizes only two dimensions, length and width, and is therefore the equivalent of using only one eye for seeing. Stereoscopic examination adds the third dimension – height. This is accomplished through the use of two overlapping photos which when viewed simultaneously creates a three-dimensional image.

Stereoscopic vision is made possible by parallax, a process defined as the apparent displacement of an object against a background when viewed from two locations. A simple exercise provides a good illustration of parallax. Begin by stretching out one arm directly in front of the body with fist clenched but the thumb raised. Observe the thumb first with the right eye, keeping the left eye closed, and then with the left eye, keeping the right eye closed. Notice how the thumb appears to shift its location against the background although, in fact, the absolute location of the thumb has not changed. This apparent shift in location represents parallactic displacement.

When an individual uses both eyes to view an object, two slightly different images are transmitted to the brain which fuses them and provides the

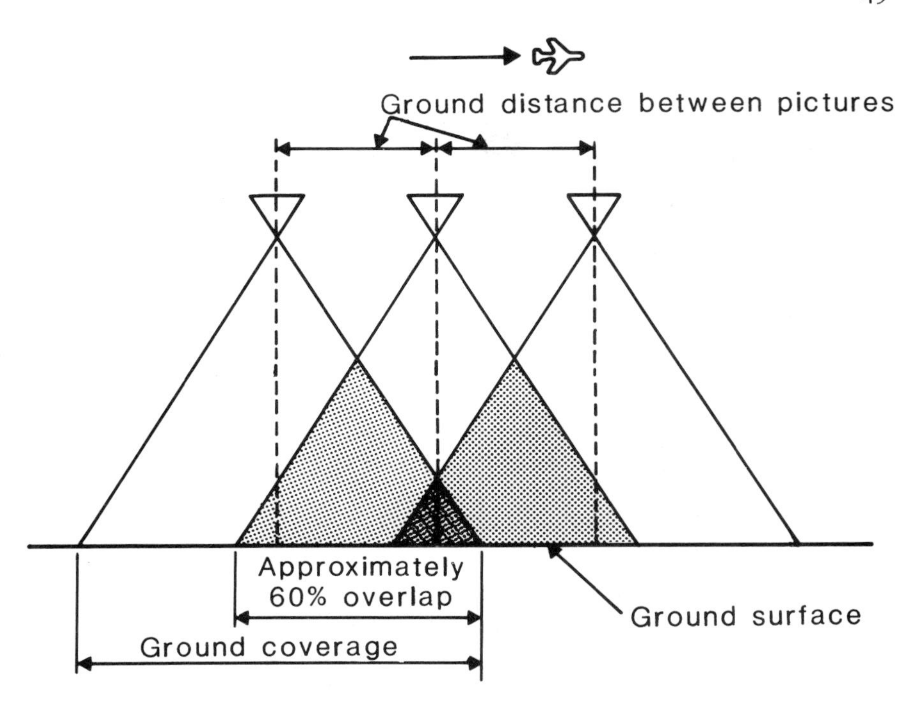

Fig. 25. Photographic coverage along flightline. Adapted from Westfall (1973).

perception of depth. There is a limit, however, to one's depth perception. The limit is about one-half mile for the unaided eye. Beyond this distance the eyes do not converge on object but remain parallel thus making stereo vision impossible. The limit to depth perception is set by the distance between the eyes (eye base). If the eyebase could be expanded, the perception of depth could occur over distances greater than one-half mile. This is exactly what happens with aerial photos. The expanded 'eyebase' represented by the distance between camera stations on two consecutive photos enables stereovision to take place at greater distances, i.e., heights.

In order to view the earth's surface in stereo, certain geometrical conditions must be met in the taking of the aerial photos. The conditions are met by taking the photos such that one photo along a flightline overlaps the next by 60 percent (Fig. 25). This is referred to as endlap. In theory, of course, 50 percent endlap would ensure an object's appearance on two photos, but the extra 10 percent is added so as to avoid gaps caused by uncompensated motions in the aircraft. And for a similar reason photos on adjacent flightlines overlap (or sidelap) on another by 30 percent (Fig. 26). Alternate photos along a flightline overlap one another by 20 percent. This is an important consideration if one is purchasing photos and has no wish to view them stereoscopically. Alternate photos provide for complete areal coverage.

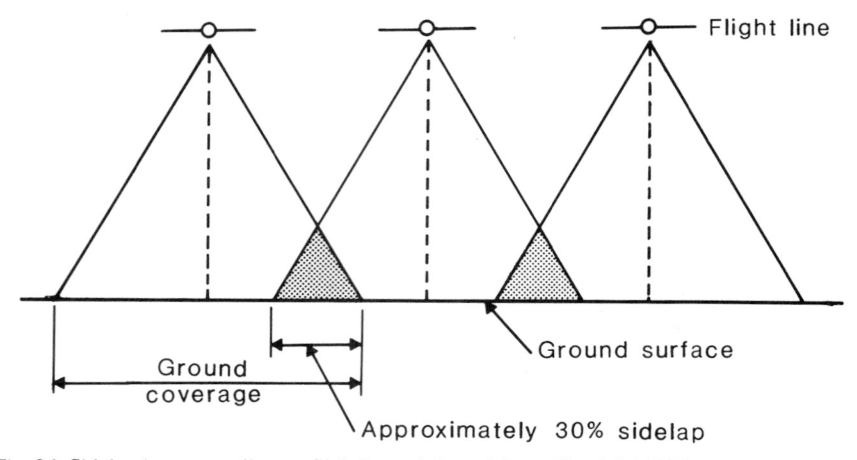

Fig. 26. Sidelap between adjacent flightlines. Adapted from Westfall (1973).

To view photos stereoscopically it is necessary to place two overlapping photos side-by-side and oriented in the same direction. A feature clearly visible on both photos should then be selected for viewing. Placing the forefinger on the left and beneath the feature on the left photo and the right forefinger beneath the feature on the right photo, the two should be drawn together until they are separated by about two and one-half inches; this distance approximates the average person's eyebase. At this point, if the left eye is focused only on the left photo and the right eye only on the right photo it is theoretically possible to see the feature three-dimensionally. However, for most people seeing stereoscopically requires the use of a stereoscope, a device that not only magnifies but also helps keep the eyes focused separately on the two images. Thus, a stereoscope is placed over the two photos with the optics spaced about two and one-half inches apart to correspond to the eyebase. When the two features are viewed through the stereoscope, they fuse towards the center providing a three-dimensional image. To sharpen the image it may be necessary to adjust the distance between the two features or the distance between the optics. Almost certainly some adjustments will have to be made before a three-dimensional image is formed.

Figures 27 and 28 are stereograms made from two overlapping photos. A lens stereoscope placed over either of these stereograms should allow the observer to view the scene three-dimensionally assuming normal or near-normal vision.

Figure 27 is an image of Ship Rock, an eroded volcanic neck located in San Juan County, New Mexico. Ship Rock provides a good test of stereo viewing because of its great height, but notice also that it is difficult to clearly view the very top. This fuzziness represents one of the effects of relief displacement described in Chapter 3.

Figure 28 is an image of Pope County, Illinois. The area is one of gently

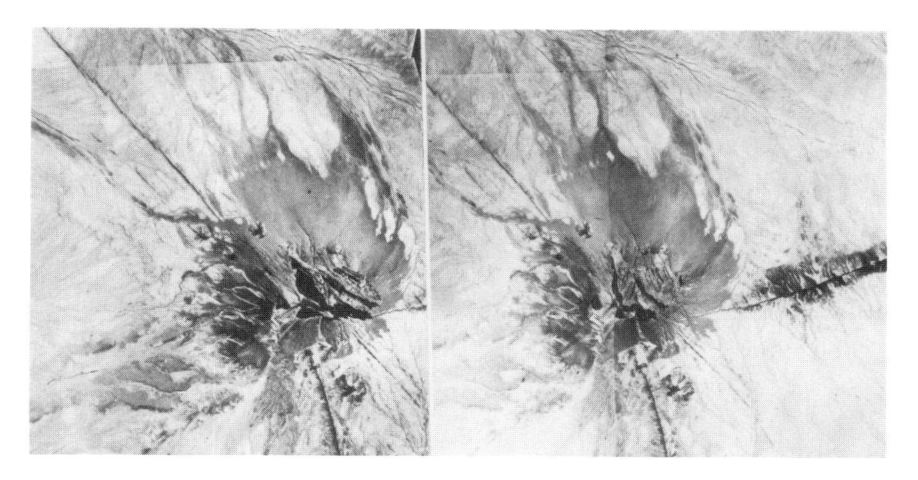

Fig. 27. Ship Rock, San Juan County, New Mexico. Scale 1:45,000. Courtesy University of Illinois, Committee on Aerial Photography.

sloping topography, the greater proportion of which is in farmland. Corn, soybeans, and hay for livestock are the most common crops and tend to occupy the more level lands. The steeper slopes have been left in woodland. Some of the wooded areas show the effects of cutting and subsequent regrowth. Over the whole area land use patterns reflect the rectangular arrangement by which land in the American Midwest was parcelled out during the 19th century.

With paper prints, particularly black-and-whites, it may be helpful to trim the margins around the photos if extensive stereo analysis is anticipated; removing the margins makes it easier to space features the required distance for stereoviewing. There may be instances, however, when even with the margins removed it may be impossible to properly space two features without bending or folding one of the prints. While paper prints can take considerable bending without serious damage, they should never be creased because the emulsion side will split; damage of this type is irreparable.

4.3 Stereoviewing Equipment

There are a number of types of stereoscopes available ranging from the simple and inexpensive to the more precision-oriented equipment of considerable expense. Three of the more commonly used types of stereoscopes are singled out here for discussion – the lens stereoscope, the mirror stereoscope, and the zoom stereoscope.

4.3.2. *Lens stereoscope.* The lens or pocket stereoscope (Fig. 29) is the least expensive of the three stereos ranging in price from about $15 to $30. It is little

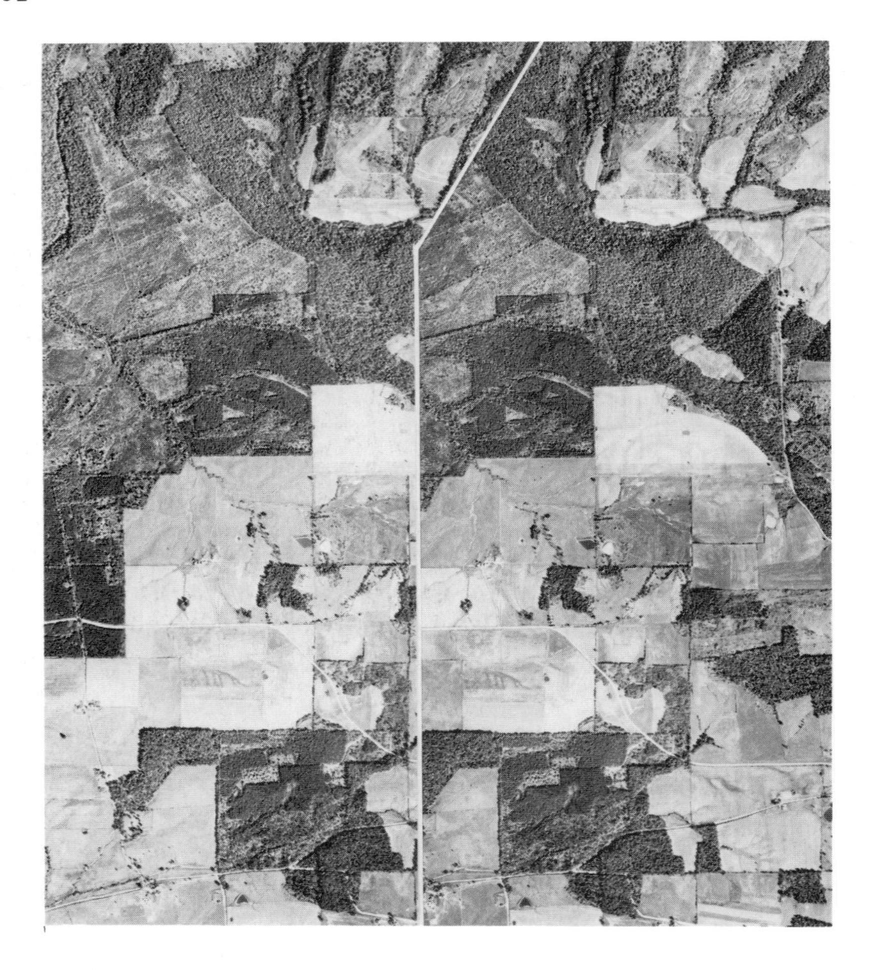

Fig. 28. Pope County, Illinois. Scale 1 : 20,000. Courtesy University of Illinois, Committee on Aerial Photography.

more than two magnifying lenses (2× or 4×) supported by a pair of thin, folding metal legs. The only adjustable part is the eyebase. The great advantage of the lens stereoscope other than its price is its size; the lens, or perhaps more appropriately, pocket stereoscope, can be conveniently folded up and taken into the field. On the other hand, its small size presents problems as well. For example, its very narrow base frequently makes it necessary to fold photos in order to see features in stereo. Furthermore, the stereoscope constantly has to be moved if any kind of annotation is being done on the photos.

4.3.3. *Mirror stereoscope.* The mirror stereoscope eliminates many of the shortcomings of the lens stereoscope but does so only at greater cost (approximately $300). The mirror stereoscope (Fig. 30) employs two sets of mirrors to

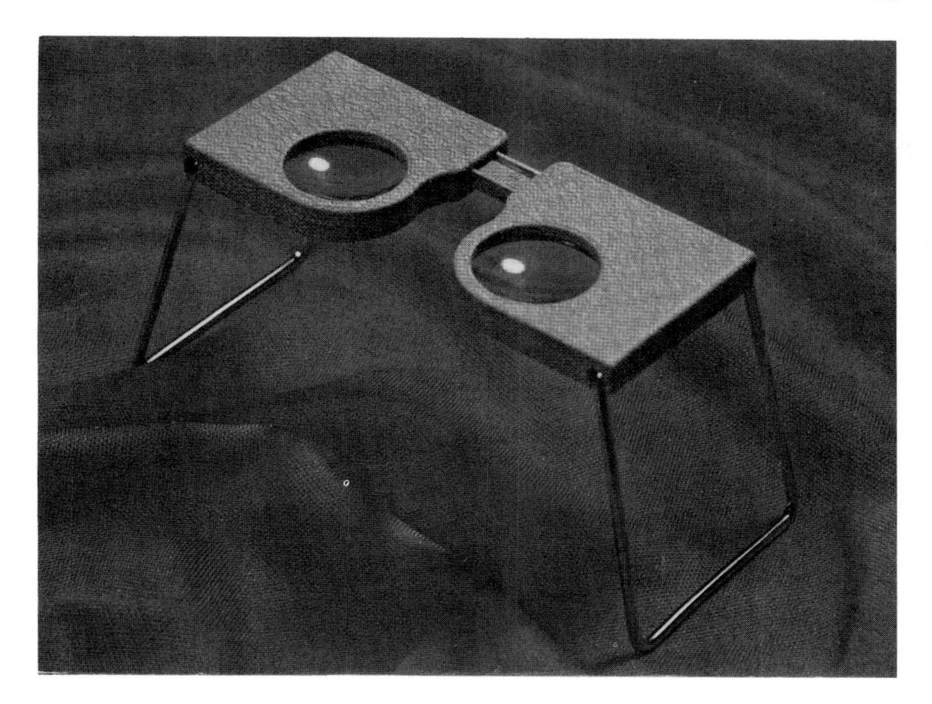

Fig. 29. Lens or pocket stereoscope. Courtesy Air Photo Supply Corporation.

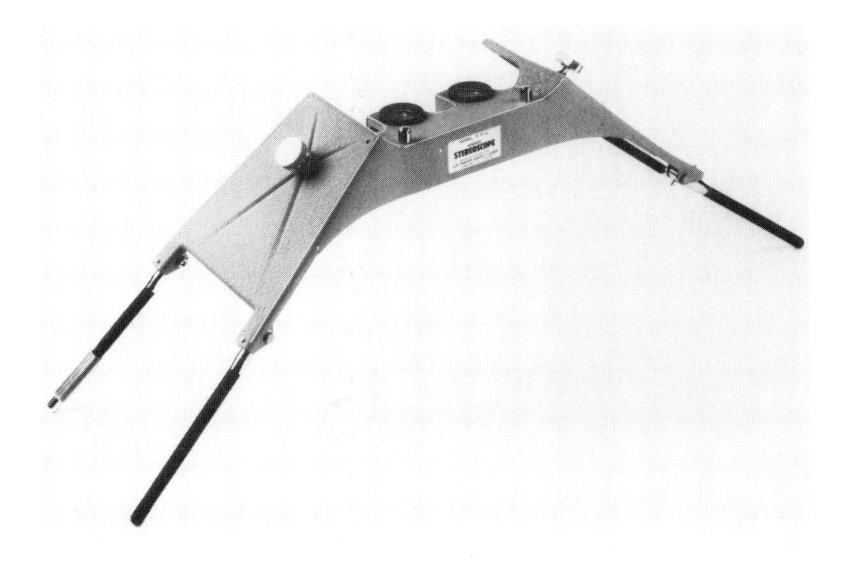

Fig. 30. Mirror stereoscope. Courtesy Forest Suppliers' Incorporated.

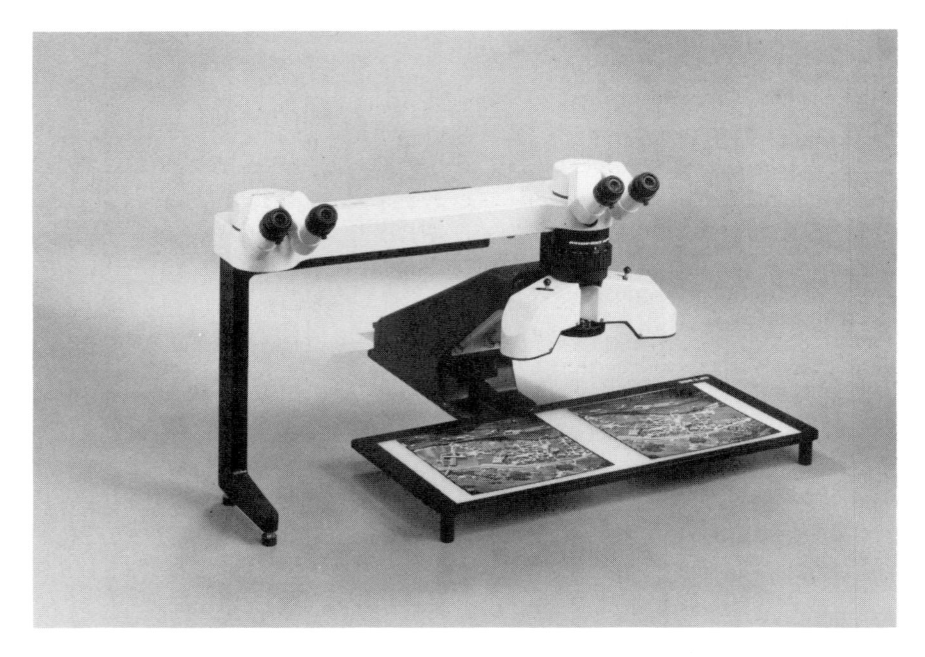

Fig. 31. Aviopret APT1. This zoom stereoscope also has a second set of optics for joint viewing of imagery. Courtesy Wild Heerbrugg Instruments, Inc.

focus the image at the optics. A set of large mirrors inclined at 45° from the horizontal reflects the image onto a smaller set of mirrors or prisms that can be viewed through the optics. The optics may either be binocular scopes or precision-ground lenses. Binocular scopes are the more expensive ($500 to $600) and have the capacity for 2 × to 4 × magnification. With lens optics, magnification can only be changed by raising or lowering them. The major advantage of the mirror stereoscope is that the two photos being examined are physically separated from one another. As a result photos do not have to be folded when viewing them stereoscopically nor does the stereoscope have to be moved in order to annotate the photos. The mirror stereoscope is too bulky to be taken into the field.

4.3.4. *Zoom stereoscopes.* The most expensive stereoscopes on the market are the zoom stereoscopes. Adding to their cost is the fact that they must be used with some type of light table. Zoom stereoscopes are precision instruments having the capability of up to 20 × magnification; 40 × when using the 2 × attachment lenses. Unlike lens and mirror stereoscopes that are generally employed for examining paper prints, zoom stereos are used almost exclusively for viewing transparencies which must be illuminated from beneath. Zoom stereos may be used for monoscopic viewing by replacing the stereo lens with a single lens. The APT1 system pictured in Figure 31 is used exclusively in stereo. It has a resolution of 100 lp/mm and a magnification of up to 31 ×.

The type of equipment appropriate to any task depends upon a number of factors – the information to be acquired, the type of photos to be used, the time available to complete the photointerpretation part of the project, whether the equipment will be used again for other projects, and perhaps most importantly the amount of money available for this purpose. The lens stereoscope in the hands of an experienced photointerpreter can be an extremely effective tool but it is not the way to conduct an investigation involving large numbers of photos. In the long run it may prove less expensive to employ color transparencies with a zoom stereoscope or a mirror stereoscope over a small light table since tests have shown that color and lighting from beneath increase interpretation accuracy.

5. How to acquire aerial photos

5.1 Introduction

There are basically three methods for acquiring aerial photos – purchase existing photos, contract to have the photography flown, and take one's own photos. The first is probably the cheapest of the three methods but one is limited to the photography available. Frequently it is impossible to obtain photos of the exact season, or film type, or scale needed. In addition, just finding what photography is available can be a time-consuming and frustrating process. These problems can largely be overcome by contracting with an aerial survey firm to have the photography flown, but there is a trade-off. While the photos obtained in this manner will undoubtedly meet the desired specifications, they will also be much more expensive than those produced from existing negatives. In a few instances, taking one's own aerial photos may provide a third alternative. Usually such instances involve the need for very up-to-date oblique or vertical photos of a rather limited local area.

5.2 Purchasing Existing Photography

The first step in deciding whether to purchase photos is to determine what photography is available. If large-scale photos are required, that is, scales of 1:10,000 and larger, it, is best to consult local officials (town or county) – engineers, tax assessors, or planning board members. Local units of government are increasingly having base maps prepared from aerial photos, and while they may not hold the photo negatives, some local official will certainly know the company that does. On the chance that local officials cannot be of help, the state is the next best source of information, in particular the state highway department. Other state agencies that may hold photography include planning offices, environmental and natural resource departments, state geological surveys, tax departments, and water resource departments.

For medium- to small-scale imagery the most efficient method for finding what's available is to utilize the new reference system established by the U.S.

Geological Survey's National Cartographic Information Center (NCIC). Inaugurated in 1975, the Aerial Photography Summary Record System (APSRS) contains records of aerial photos acquired by a number of federal agencies including the U.S. Geological Survey, NASA, the National Ocean Survey, the Environmental Protection Agency, the Department of Agriculture, the Department of Defense, and the Tennessee Valley Authority. All photographs listed in the APSRS files are available for purchase from the contributing agencies.

The basic unit of information in the APSRS system is the summary record which describes specific aerial photographic projects. The information contained in the record includes the holder of the photographs, the date of the photos, the film used, scale, amount of cloud cover and so on. APSRS does not hold the actual photographs, however, only information about them.

The summary records may be produced for users in any of three formats: APSRS graphics (maps locating the records), microfiche, and customer queries. The graphics and microfiche must be used directly. They are grouped by state and quickly reveal the most recent imagery available. The customer query is for complex requests. It is fed into NCIC's computer with the resulting printout furnished to the requester. Information about the APSRS may be obtained from NCIC national headquarters:

National Cartographic Information Center
U.S. Geological Survey
507 National Center
Reston, Virginia 22092
Telephone: (703) 860-6045

A program of particular note is the National High Altitude Photography (NHAP) program. Begun in 1980 the program is intended to acquire complete aerial coverage of the conterminous United States every several years. Approximately one-half million square miles of the U.S. are photographed each year with completion of the first-time coverage expected by 1986. At present more than 100,000 high quality black-and-white and color-infrared photos have been taken as part of this program.

Both the black-and-white and the color-infrared photos are taken at altitudes of 40,000 feet above sea level. The black-and-white photos have a scale of 1:80,000 and each covers approximately 130 square miles. The color-infrared photos have a scale of 1:58,000; each of them covers approximately 68 square miles. All photos are high-resolution and may be viewed as stereo pairs. They may be ordered in either a 9 x 9 inch format or as enlargements in sizes ranging up to 36 inches square. Information about this program may also be obtained from the National Cartographic Information Center.

On occasion there may be a need for old photographs to establish a base for analyzing land use change or the impact of a large engineering project (flood-control dam or interstate highway) on the local environment. Aerial photos of the united States taken by federal agencies before 1941 have been assembled by the Center for Cartographic and Architectural Archives at the National Archives. These photos date from the mid-1930's and cover approximately 80 percent of the continental United States. On request the Archives will search the files and return a research report and a price list for prints. Inquiries should be addressed to:

National Archives
8th and Pensylvania Avenue, N.W.
Washington, D.C. 20408
Telephone: (202) 523-3006

As a general rule, photos purchased from the Federal Government tend to be less expensive than those purchased from private companies. This is particularly true if only a few prints are to be ordered. When ordering large numbers of prints, however, the price difference becomes markedly less. Table 5 shows the most recent prices for U.S. Department of Agriculture photos.

5.3 Contracting for Photographic Mission

The objective of any aerial photographic mission is the acquisition of high quality photos from which needed information can be extracted. To accomplish this objective, however, both careful planning in advance and precise execution of the mission itself are required. For this reason, then, it is recommended that such a project be undertaken only by an experienced and reputable aerial survey company.

Mission planning begins with the specification of the problem, that is, what information is to be extracted from the photos. Will it be numbers of residences, location and acreage of wetlands, amount and kinds of new construction? Whatever the information, it will determine many important aspects of the mission including the type of film and filter to be used, the optimum scale of the photos, and the optimum time of day and year for the photography to be taken.

5.3.2. *Film and Filter Combination*. The appropriate film and filter combination to use may depend on several considerations. Cost is one. Black-and-white film is considerably less expensive to use than color films and is more reliable. On the other hand, black-and-white film may not be the most effective in terms of the data to be extracted. For example, color film is better than black-and-white for water quality analysis, while color-infrared is usually the best for vegetation mapping. At times cost may not even be an important factor. For high altitude flights where the cost of the aircraft alone may run to $1500 per hour, film and processing costs may represent as little as 5 percent of the total.

For many tasks the decision on which film and filter to use is a relatively

Table 5. Prices of USDA aerial photos*

Type and Size of Reproductions	Black-and-White		Color (National Forest Areas Only)			
			Color Negative		*Color Positive*	
	Paper	*Film Base Pos. Transp.*	Paper	Film Base Pos. Transp.	Paper Reversal Prints	Film Base Pos. Transp.
Photo Indexes						
20″ × 24″	$ 5.00	$15.00	N/A	N/A	N/A	N/A
Contact Prints						
10″ × 10″	2.00	3.00	$ 5.00	$12.00	$ 7.00	$12.00
Enlargements						
12″ × 12″	6.00	10.00	15.00	--	N/A	20.00
17″ × 17″	7.00	12.00	N/A	Prices	N/A	N/A
20″ × 20″	N/A	N/A	20.00	on	35.00	35.00
24″ × 24″	8.00	15.00	25.00	Request	35.00	35.00
38″ × 38″	25.00	30.00	40.00	--	45.00	60.00

*Prices as of May, 1984.

simple one; on occasion, however, the nature of the project may require careful evaluation of several possible combinations. For this purpose, the Photographic Systems Simulation Model has been developed. The model predicts the contrast in image tone to be expected between a feature and its background for various film/filter combinations. The only inputs the user must have are spectral reflectance curves for the feature and background of interest. A spectrum of atmospheric conditions, sensitivity characteristics for commonly used films, and transmission characteristics for various filters are on file for use in the program (1).

5.3.3. *Scale*. As described in Chapter 3, scale is a function of focal length and flying height above the terrain and can be computed as follows:

$$\text{Scale (RF)} = \frac{\text{fl}}{\text{H-h}}$$

where fl = focal length
H = flying height above sea level
h = average terrain height above sea level.

As a rule it is best to select the smallest scale that will still allow the smallest unit of information needed to be discernible on the photography. National Map Accuracy Standards require that planimetric features be plotted to within 1/30 of an inch (about 0.85 mm). Thus, in order to attain an accuracy of ± 5 feet (1.6 m), a map scale of 1 : 18,000 is necessary. At this scale 1/30 of an inch on the map or photography will represent about 5 feet on the ground. This is one approach for determining a suitable scale photography for a thematic mapping project (1).

There are several standard focal lengths available for photographic missions – the $3\frac{1}{2}$ inch (88 mm), 6 inch (152 mm), $8\frac{1}{4}$ inch (209 mm), and 12 inch (304 mm). The 6-inch focal length is the standard for most mapping work. Longer focal lengths are used in areas of high relief to minimize displacement effects, while shorter focal lengths are used over areas of little relief. And, since scale is inversely proportional to focal length, shorter focal lengths cover a larger ground area than longer focal lengths. If two scales of photos are needed of an area, it may only be necessary to fit out two camera systems with difference focal lengths to obtain them.

Scale is also determined by flying height, the greater the altitude of the aircraft the smaller the scale. Missions are generally flown at high altitudes when there is a need to minimize the effects of relief. For topographic mapping, however, missions are flown at low altitudes to take advantage of the increased effects of relief displacement. Unfortunately, distortions caused by scale variations and tilt displacement also increase at lower altitudes.

In summary, then, smaller scales require fewer photos to cover a given area.

This results in fewer photos to interpret and a possible savings in cost. One note of caution should be added here, however. The savings in cost realized from a decrease in photo scale applies only to lower altitude flights. High altitude flights are expensive with costs increasing geometrically with altitude. These higher costs result from the need for larger aircraft, increased maintenance, and for more ideal flying conditions.

5.3.4. *Time.* The scheduling of an overflight may involve considerations as to time of year, time of day, and possibly even day of the week. Again, it depends upon the information needed. Time of year is critical for vegetation or wildlife habitat mapping. Photos for such purposes are acquired during the growing season when deciduous vegetation in particular has reached maturity. Agronomists may need photos on several occasions both for monitoring growth and disease detection. On the other hand, for census mapping or topographic mapping it is better to acquire photos in the spring or fall when there are fewer leaves on the trees to mask the terrain. Unfortunately, spring and fall are two seasons when the weather is not dependable. Long layovers waiting for suitable flying weather can add appreciably to the cost of a flight.

Time of day considerations are usually involved with sun angle or the need to minimize the effect of shadows. To reduce the effect of shadows, photos are acquired within one to two hours of local noon. However, shadows are also an important aid to identification so they should not necessarily be eliminated entirely. There are other time of day considerations. For example, traffic engineers may wish photos of 'rush-hour' traffic so they should be obtained between either 7 a.m. and 9 a.m., or between 4 p.m. and 6 p.m. For this same reason day of the week may be important. Rush-hour implies a week day, but other critical times for traffic may be holidays or summer weekends. Weekend photos have advantages of their own. In one instance, photos taken on a Saturday made it easier to differentiate suburban shopping malls from high-tech industrial parks. Though architecturally similar, shopping malls had full parking areas, high-tech industrial parks relatively empty ones.

5.3.5. *Flight Plan.* Once the desired camera format and photo scale have been determined a flightmap must be prepared as a guide for the pilot and the photographer. Though usually drawn up by the survey company itself, it is useful for the photointerpreter to participate in the process. On a map of the area to be photographed a series of flightlines are laid out parallel to each other and to the long dimension of the project area. If the mission is designed to provide stereophotos, then the proper 30 percent sidelap between adjacent flightlines and the 60 percent endlap for adjacent photos along the flightline must be precisely calculated as a function of altitude and airspeed. These calculations are most readily made from tables established for this purpose.

The success of any aerial photographic mission depends heavily upon the skills of the pilot and photographer. The pilot, using predetermined landmarks

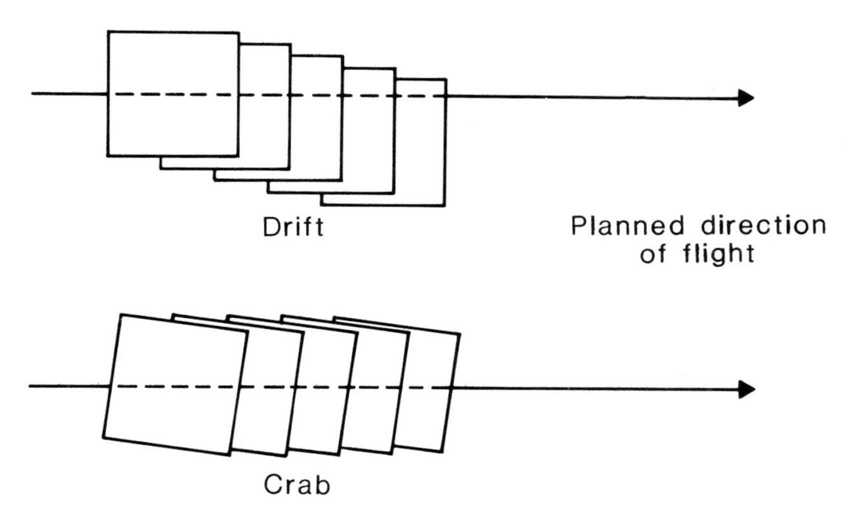

Fig. 32. Schematic diagram illustrating crabbing and drifting.

as a guide, must carefully duplicate the pattern of flightlines laid out on a map. The aircraft must be flown straight, level, and at the correct altitude. At the same time the photographer must see to it that the camera mount is maintained in a level position and the flightline is being followed precisely. The camera is usually shuttered automatically by a device called an intervalometer. If a crosswind is causing the aircraft to drift off its course, the photographer must instruct the pilot to head the aircraft into the crosswind. This technique, known as crabbing (Fig. 32), will compensate for aircraft drift. However, crabbing (and drifting) may reduce overlap to an unacceptable level causing the photos to be skewed. Thus, the photographer must also compensate by changing the orientation of the camera; the sides of photos must be parallel to the principal point base line. For both pilot and photographer low-altitude flights are more difficult than medium- to high-altitude flights. The horizon is not visible so tracking must be done through the viewfinder leaving little time to correct errors. Atmospheric turbulence may compound these difficulties even further.

5.3.6. *Technical Specifications.* Flight plans include a detailed set of specifications usually set forth in a formal contract. The contract normally lists the materials and equipment to be used, the procedures to be followed, and the final products to be delivered. It also details the requirements and tolerances pertaining to scale, overlap, tilt, and image quality.

When a photographic survey had been completed it is in the best interest of the purchaser to carefully evaluate the final product. This means checking to see that tolerances have not been exceeded. All prints should be properly dated in the upper left hand corner and display the project, flightline, and exposure numbers in the upper right corner. At least a number of prints should be checked for scale. Distances between points on the photos can be compared to distances computed between the same points on the flight map. Allowing for variations on relief, average scale should be within 5 percent of the specified scale. Photos should also be checked for excessive tilt. Excessive tilt is generally self-evident. Finally, there is the question of print quality. Are photos blurred or washed-out looking? If negatives are of reasonable quality there is no reason why final prints should not be of good quality. Survey companies have every reason to see that the purchaser is satisfied with the final product.

5.4 Small-Format Aerial Photography

When the costs of conventional aerial photography preclude its use, taking one's own aerial photos may be a feasible option. The most popular formats for such missions are 35 mm and 70 mm due to the low cost and easy availability of both cameras and film. The main disadvantages of these two formats are the small field of coverage and the lack of calibration. Compared to standard 9 x 9 aerial photos, it takes many more 35 mm or 70 mm photos to cover the same ground area at the same scale. As for the lack of calibration, small-format cameras have no mechanism for holding the film flat or for producing fiducial marks. While there are methods for making geometric corrections on small-format photos, in practice most applications do not require such accuracy [14].

Oblique photos for monitoring changes in land use can generally be acquired with a minimum of advanced planning. For information on this topic Fleming and Dixon's *Basic Guide to Small Format Hand-Held Oblique Aerial Photos* [15] is highly recommended. The acquisition of small-format vertical photos, on the other hand, requires considerable preflight preparation. Almost any light plane can be fitted out to carry a small-format camera system although high-winged aircraft are preferred because of their superior downward visibility. For the best quality photos the camera should be mounted in some way. Depending upon the configuration of the aircraft, the camera could be mounted in the floor, on the window, or on the side of the aircraft. The latter is the least flexible. During flight, exposures must be made at regular intervals by means of a remote cable shutter release and a motor drive unless the camera is electrically operated in which case rechargable batteries within the camera power the film advance, shutter cocking and release. Side mounts may also make impossible the changing of film or camera settings in flight. This may limit the mission to the number of exposures on a single roll of film. One advantage to using 70 mm film is that a 100-foot roll contains over 450 exposures.

Even a small mission requires the preparation of a flight plan. The pilot, for example, must know what altitude and airspeed is to be maintained as well as what route is to be followed. The photographer must know how many photos are to be taken and what the interval is to be between exposures. The altitude

Fig. 33. Oblique 35 mm photo of White River Junction, Vermont. May 1982. Courtesy Dan Baum.

is computed as a function of the required photo scale and the focal length of the camera (see Table 6). The number of exposures needed to cover the project area can be determined with the aid of a neat model. A neat model shows the ground area to be covered by a single exposure at a given scale. Thus, a number of neat models superimposed over a map of the study area will show the number of exposures and flightlines needed to cover the area with the required overlap. The exposure interval is computed as a function of aircraft speed and the distance between the principal points of the planned exposures.

The low altitudes at which small-format photos are acquired (often under 1,000 feet) make weather an important consideration. Small aircraft can be severely affected by turbulence and local winds and there is little margin for error. On the other hand, the effects of haze are almost nonexistent at low altitudes and even clouds are less of a limiting factor. It has been shown that excellent quality, shadowless photos can be obtained under overcast skies as long as there is no rain and there is sufficient light to expose the film [16]. The type of film to be used on a small-format photographic mission frequently depends upon whether the final product desired is to be in the form of paper prints or slides. If the final product is to be prints, panchromatic or color-negative films are recommended; prints made directly from negatives are usually

Table 6. Coverage table for 35 mm aerial photography* (24 mm × 36 mm format)

	Altitude above ground datum			Photo coverage		
	Camera focal length			Length	Width	Area
Scale	35 mm	55 mm	100 mm	(ft)	(ft)	(acres)
1:1200	150	220	390	90	140	0.3
1:2400	300	430	790	190	280	1.2
1:3600	450	650	1180	280	420	2.8
1:4800	590	870	1580	380	570	4.9
1:6000	740	1080	1770	420	640	6.2
1:7200	890	1300	2360	570	850	11.0
1:9600	1190	1740	3150	750	1130	19.6
1:12,000	1480	2170	3940	940	1420	31.0
1:20,000	2480	3610	6570	1410	2120	68.8
1:36,000	4460	6500	11,820	2830	4250	276.0
1:48,000	5940	8670	15,760	3770	5670	490.0

*Adapted from Merle Meyer, *Operating Manual – Montana 35 mm Aerial Photography System*, Research Report 73–3, University of Minnesota, 1973, p. 23.

less expensive and of better quality than those made from slides. The latter are typically made from color-reversal films one of which is color-infrared.

The use of color-infrared film for small-format missions requires some caution. For one thing only one type of infrared film, Ektachrome Infrared (Type 2236), can be purchased over the counter at camera supply stores and it comes only in rolls of 35 exposures. The more commonly used infrared film, Aerochrome Infrared (Type 2443), must be ordered from the manufacturer and in bulk lots [17]. For another, color-infrared films require refrigeration both prior to and after exposure. To make certain that film rolls have been properly refrigerated, it is best to shoot a test roll prior to the main mission. After a mission exposed rolls of infrared film should be packed in dry ice until ready for processing. A final note of caution involves the processing itself. Since few color labs have had much experience in processing color-infrared film, care should be taken to find one that has.

Since the cost factor has been expressed as one of the important reasons for turning to small-format systems some actual costs figures are in order. Watson has presented cost figures for six operational projects conducted in Canada [16]. On the basis of these projects one could estimate the cost of acquiring and processing 35 mm photos at about $1500 depending upon the size of the area covered and the scale of the photos. Most of the costs (aircraft rental, film and processing) appear to be fairly standard. Labor is the one area that can vary and where some savings might be realized.

In summary, small-format aerial photography provides an effective alternative to conventional aerial photography. It is usually less expensive, can be acquired on a timely basis, and can be processed rapidly after each flight.

6. The Landsat system

6.1 Introduction

To this point the discussion has focused entirely on conventional aerial photo-interpretation techniques and for good reason – the bulk of remote sensing data used by land planners is acquired from conventional aerial photos. There is one other system whose products have proved to be of value, particularly when information is needed over large areas. Landsat is an earth-orbiting satellite system which acquires multispectral data of the earth's surface on a continuous basis. Since the program's inception in 1972, Landsat (actually five separate satellites) has acquired the equivalent of millions of frames of information and continues to do so. Landsat data have already become important for geologic exploration, crop production estimates and rangeland management.

6.2 Landsat System

Landsat-1 was launched on 23 July, 1972, the first satellite ever to be placed in orbit for the specific purpose of acquiring continuous multispectral data of the earth's resources. At the time of launch, Landsat-1 was known as ERTS-1 for Earth Resource Technology Satellite. In 1975, NASA officially changed the name of the program to Landsat, and subsequent satellites have been designated Landsat-2 (1975), Landsat-3 (1978), Landsat-4 (1982), and Landsat-5 (1984). The first two Landsat satellites were identical. Both have now been terminated after several years of service. Landsat-3, a slightly modified version of Landsats -1 and -2, has also been terminated. Landsat-4 is presently operational but it is experiencing technical difficulties and will be used only for special acquisitions; Landsat-5 will assume the main data acquisition load. Because Landsat-4 and -5 have significantly different sensor systems and orbital characteristics, they will be described separately.

Landsats-1, -2, and -3, hereafter referred to simply as Landsat, were launched into identical near-polar orbits at a nominal altitude of 917 km (570 miles). The orbit of a Landsat satellite is sun synchronous, meaning that the orbital plane of

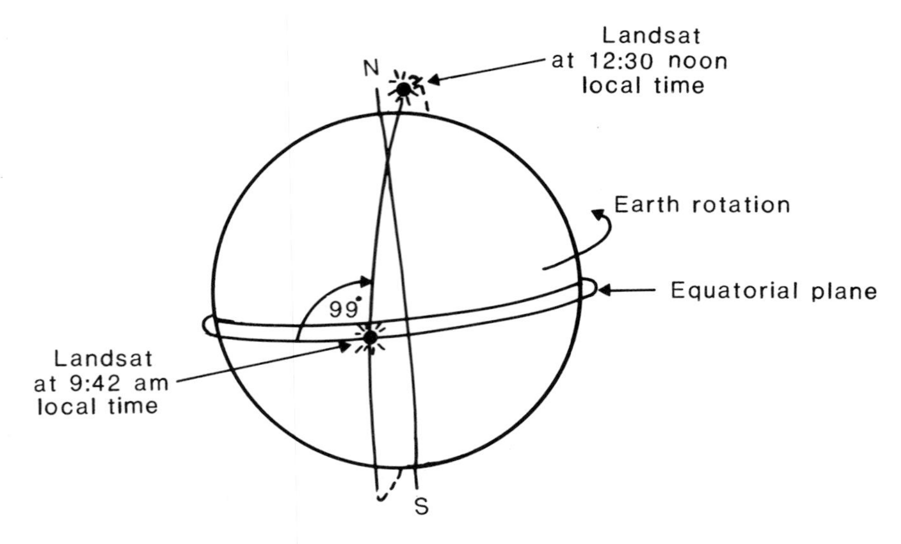

Fig. 34. Inclination of Landsat orbit. Adapted from NASA.

the satellite precesses about the earth at the same angular rate as the earth revolves about the sun; the orbital plane is inclined 99° from the equator measured in clockwise direction. Thus Landsat crosses the equator at the same local time (9:30 a.m. to 10:00 a.m.) on each pass over the sunlit side of the earth (Fig. 34). This insures that adjacent areas may be seen on successive days with no detectable change in sun angle or direction; this is an important consideration in the construction of mosaics which require the use of imagery taken days apart. The sun synchronous concept should not be misinterpreted. Sun angle is not a constant; it varies with latitude and time of year. In January, for example, sun angle may vary from 0° near the North Pole, where the sun may not rise above the horizon, to nearly 45° at the equator.

Passing north to south over the sunlit side of the earth, Landsat completes one orbit in 103 minutes. On the very next orbit Landsat covers a track well to the west (2875 km/1785 miles to the west at the equator) as the earth rotates beneath the spacecraft. It is the next day, or 14 orbits later, before Landsat completes the pass adjacent to orbit one; its location will be 159 km (99 miles) west of orbit one at the equator (Fig. 35). For complete coverage of the earth it requires 18 days or 251 orbits; orbit 252 begins a new sequence by duplicating orbit one. When Landsat-2 was launched, Landsat-1 was still operating, so by placing Landsat-2 in orbit 9 days behind Landsat-1 it was possible to provide complete earth coverage on a 9-day cycle.

Like aerial photos, Landsat imagery contains overlap because the orbits converge near the poles. At 81° north and south the amount of overlap (sidelap) is at a maximum – about 85% (97 km/61 miles), but decreases from there to about 14% (26 km/16 miles) at the equator (Table 7). At the poles then, it is

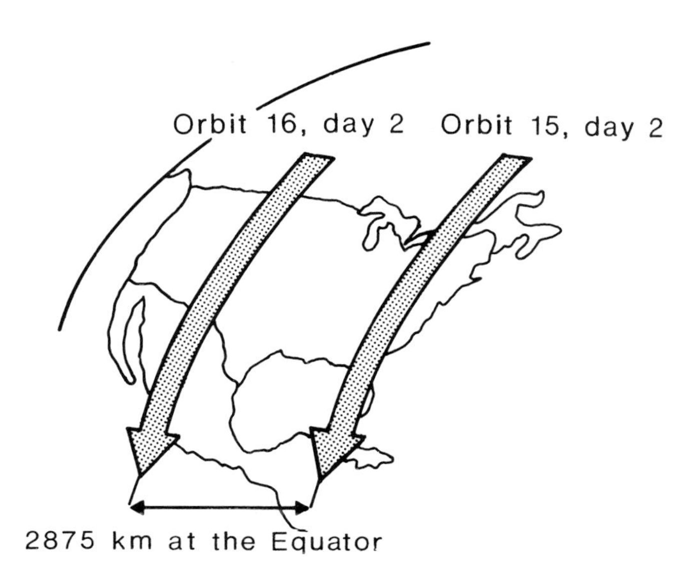

Fig. 35. Landsat ground coverage pattern. Adapted from Short (1982).

Table 7. Landsat image sidelap by latitude of acquisition

Latitude (degrees)	Image sidelap (percent)
0	14
10	15
20	19
30	25
40	34
50	45
60	57
70	70
80	85

possible to have day-to-day coverage barring cloudy weather. Overlap also makes possible the viewing of Landsat imagery stereoscopically. In practice, however, little stereo-viewing is done, because the high altitude of the spacecraft keeps vertical exaggeration to a minimum. For this reason, Landsat imagery is nearly orthographic.

6.2.2. *Landsats-1 and -2*. The instrumentation aboard Landsats-1 and -2 was identical. It included a return beam vidicon (RBV), a multispectral scanner (MSS), a data collection system receiver and transmitter (DCS), and two wide band video tape recorders (WBVRT's). The RBV system consisted of three television-like cameras that were to have acquired data simultaneously in three spectral bands. Unfortunately, technical problems with the RBV system on Landsat-1 forced its shutdown after producing only about 1700 images. The bulk of the data available from Landsat, therefore, are those acquired by the MSS system which continues to operate on Landsats-4 and -5. For this reason, much of the discussion in this chapter will focus on the MSS. The DCS, of which little will be said, is a real-time data relay system which receives signals from a number of ground based data collection platforms and transmits them to NASA's Goddard Space Flight Center in Greenbelt, Maryland. The platforms, placed in remote locations, collect primarily hydrologic data such as air temperature, stream flow, and snow depth. The WBVTRs, found only on Landsats-1, -2, and -3, are used to record MSS and RBV data for later transmission when the spacecraft is not within line-of-sight of a ground receiving station. NASA maintains three receiving stations in the United States – at Goddard, in Goldstone, California, and in Fairbanks, Alaska. A number of other countries have also built receiving stations, including Canada, Italy, Sweden, Japan, India and Australia.

The most important instrument aboard Landsat spacecraft to this point has been the MSS. The MSS scans the earth's surface by way of an oscillating mirror as the satellite passes southward (Fig. 36). Radiation reflected from the earth is directed by the mirror to a set of fiber optics which in turn conduct the radiation to a series of 24 detectors. Filters allow only the desired spectral data to pass through. The radiation reaching the detectors causes them to produce an electrical charge (between 0 and 5 volts) proportional to the amount of radiation received. These data which are in analog form are converted to digital values

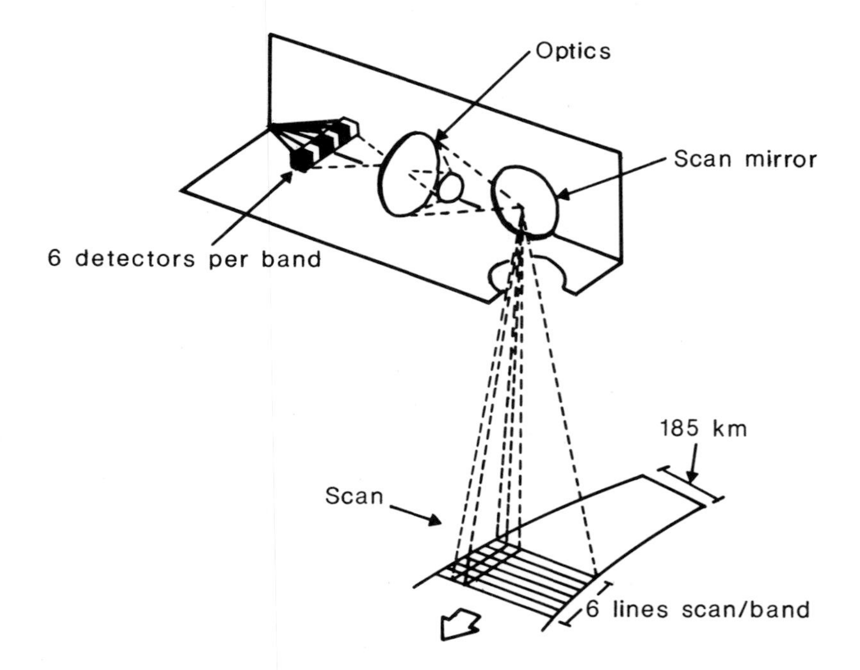

Fig. 36. Operation of Landsat MSS. Adapted from Short (1982).

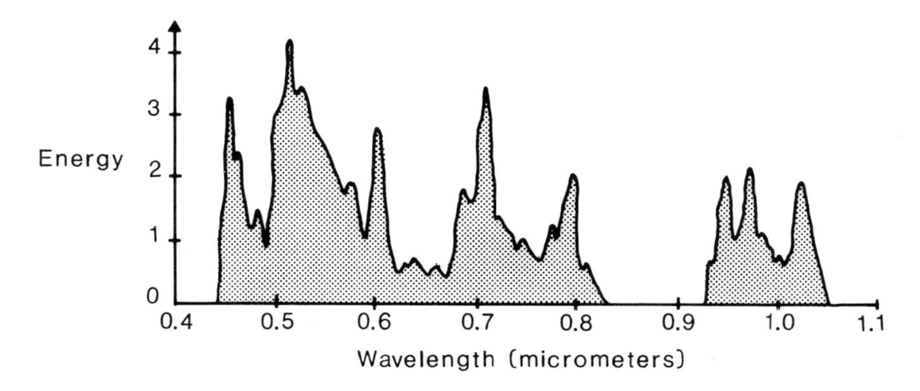

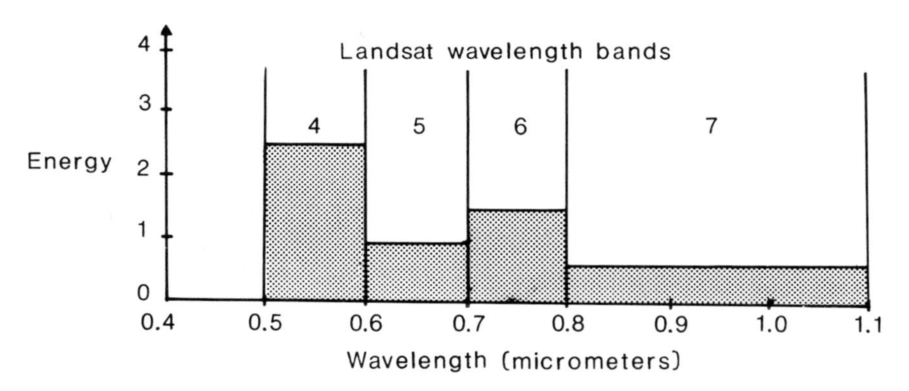

Fig. 37. Spectral signature as acquired by Landsat MSS. Adapted from Department of the Army (1979).

(between 0 and 63) and either transmitted directly to a ground receiving station or recorded on one of the WBVTR's for later transmission. These transmitted data represent the amount of radiation (the brightness) reflected from a given area of the earth's surface.

The MSS scans west to east across the spacecraft's orbital path. Each scan covers a ground distance of 185 km (115 miles). Radiation is recorded in four spectral bands – one green, one red, and two reflective infrared (Fig. 37). For each band there are six detectors, so a total of 24 lines of data are acquired simultaneously. While the scanning takes place on a continuous basis, the data are framed to present an image covering an area 185 km × 185 km with a 10 percent overlap between adjacent frames. A nominal MSS scene consists of 2340 scan lines.

Data are generated by sampling the voltage output of each detector at set intervals – approximately every 10 microseconds. For each scan line this represents 3300 samples and with 2340 scan lines, it means more than 7.5 million

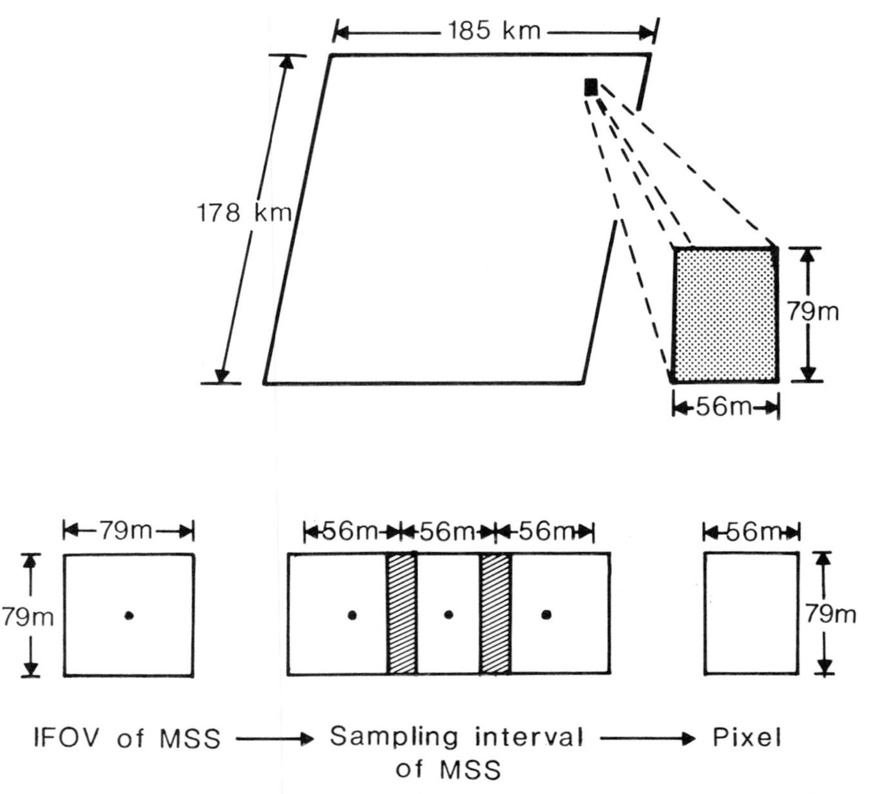

Fig. 38. Dimensions of pixels on standard Landsat MSS image. Adapted from NASA.

data points are found on each MSS scene. The ground dimensions of each sample are determined by the scanner's instantaneous field of view (IFOV) which is a square area measuring 79 m x 79 m on the ground. Thus, each sample is a measure of the total radiation reflected in each of four spectral bands from a 79 m x 79 m area of the earth's surface (Fig. 38). However, the 10 microsecond interval allows the scanner to move only the equivalent of 56 m on the ground, an overlap of 23 m. Therefore, to reduce the geometrical inaccuracies arising from this situation, the spectral measurements made from the 79 m x 79 m area are formatted as if from an area 79 m x 56 m. This latter area is referred to as a Landsat picture element or 'pixel'. The pixel is the smallest unit for which data are collected and as such determines ground or spatial resolution.

6.2.3. *Landsat-3*. The Landsat-3 system differs from its predecessors in two important respects. The first relates to the MSS system. On Landsats-1 and -2 the MSS acquired data in four spectral bands – green, red, and two reflective infrared; on Landsat-3 an additional channel has been added in order to detect thermal radiation emitted from the earth's surface at 10.2 to 12.6 μm. Once again, however, technical difficulties have led to the premature shutdown of this channel, so few thermal data are available to analysis.

The second, and more significant, difference in Landsat-3 is the result of modifications in the RBV system. The RBV system on Landsats-1 and -2 utilized a three-television camera system to simultaneously acquire data in three spectral bands. No film is used. Instead, reflected radiation is focused on a phosphor-coated tube with each shuttering of the camera. The phosphor-coated tube is scanned by an electron beam to detect minute voltages. These are amplified, converted to digital values, and transmitted to one of the ground receiving stations. As with the MSS system, RBV data are stored on one of the WBVTR's when the system is not in direct contact with a ground station. On Landsat-3 the three-television camera system has been reduced to a two-camera system which is shuttered in sequence rather than simultaneously. Thus, the electron beam has only one screen to scan at a time rather than three, and resolution is vastly improved. The two cameras also acquire data in the same spectral band, a broad one ranging from 0.5 μm to 0.75 μm. They are aligned to view adjacent 98 km x 98 km areas with a sidelap of 13 km. Two successive pairs (or four scenes) cover the same area as one MSS scene; they are designated A, B, C and D. One final improvement has been the doubling of the focal length of the camera optics. The result is resolution approaching 30 m, a figure more than double that of the 80 m MSS system.

From the beginning it was anticipated that the RBV system would be superior to the MSS system. For one thing, the RBV system images an entire scend simultaneously rather than line-by-line. This provides RBV data with greater cartographic fidelity than MSS data. Furthermore, the RBV system contains a grid in the image plane that facilitates making geometric corrections. Unfortunately, though the RBV system on Landsat-3 had been providing high quality data, the early failures shifted the emphasis of the Landsat Program to the MSS. The RBV system terminated with Landsat-3, as it has been replaced by a new system on Landsats-4 and -5.

6.2.4. *Landsats-4 and -5*. The first three Landsat satellites were experimental and contained essentially the same sensor package. Landsats-4 and -5, however, are officially considered operational, second-generation systems. As such, they have undergone considerable modifications. These modifications have provided Landsats-4 and -5 with improved spatial resolution, greater spectral separation, and higher radiometric accuracy in comparison with the earlier systems. They have also been designed to interact with a number of other satellite systems for data relay, communications, and orbit control.

To improve spatial resolution, a critically important consideration for land planners, Landsats-4 and -5 have been placed in a lower orbit. While still a sun synchronous, near polar orbit, they have a nominal altitude of 705 km (438 miles) compared to 920 km (570 miles) for Landsats-1, -2, and -3. Landsats-4 and -5 orbit the earth every 100 minutes, crossing the equator between 9:30 a.m. and 10:30 a.m. local time, and complete 14.5 orbits per day. Complete

Table 8. Thematic Mapper spectral characteristics*

Band	Spectral range (μm)	Principal applications
1	0.45 to 0.52	Coastal water mapping Soil/vegetation differentiation
2	0.52 to 0.60	Green reflectance by healthy vegetation
3	0.63 to 0.69	Chlorophyl absorption for plant species differentiation
4	0.76 to 0.90	Biomass surveys Water body delineation
5	1.55 to 1.75	Vegetation moisture measurement Snow/Cloud differentiation
6	10.4 to 12.5	Plant heat stress Other therman mapping
7	2.08 to 2.35	Hydrothermal mapping

*General Electric, *Landsat-D*, SSD 1281–20, p. 5.

coverage of the earth requires 16 days (233 orbits) rather than the previous 18 (251 orbits). The lower orbit also changes the coverage cycle. For Landsats-4 and -5 the adjacent swath to the west of a previous one takes place not on the next day but seven days later. The distance between any two consecutive swaths is 2752 km (at the equator). The distance between the swaths is filled in over the 16 day period. As for sidelap, it is least at the equator (7 percent), increasing in the direction of the poles.

Greater spectral separation and higher radiometric accuracy have been attained by replacing the RBV system with a new one called the Thematic Mapper (TM). The TM operates in seven spectral bands (Table 8), six of which were selected primarily for vegetation monitoring, the seventh for its ability to discriminate between rock types. As the scan mirror sweeps back and forth along the same 185 km swath, data are collected in both directions, unlike previous Landsat systems which only acquired data in one direction. A raster array of 16 detectors for each of bands 1–5 and 7, and four for band 6 (thermal) translates reflected radiation into electronic signals which are amplified, converted to digital form and transmitted to ground receiving stations. To improve radiometric accuracy of the data the range of values has been increased from 0–64 to 0–256.

The lower orbit of Landsats-4 and -5 enables the TM to have higher spatial resolution than previous systems. A pixel size of 30 m × 30 m in all but band 6 will allow for the classification of areas as small as 2–4 hectares (approximately 6–10 acres); band 6 has a pixel size of 120 m × 120 m. The number of pixels per full resolution image (again excluding band 6) is about 42 million. Unfortunately, the technical problems with Landsat-4 have severely limited data acquisition by the TM although a few scenes are commercially available (Fig. 39). Additional TM data will have to await for the operation of Landsat-5.

Fig. 39. Band 1 (blue) Thematic Mapper image of Saginaw Bay, Michigan. Clearly visible are the cities of Saginaw and Bay City. Also visible is the line demarcating the heavily farmed lake plain from the sandy soils and less heavily farmed area representing the morainic uplands. Courtesy NASA.

MSS data from Landsat-4 are available. However, the lower orbit of Landsat-4 required some modification to the MSS system. Basically, this modification involved adjusting the optics so the pixel size of Landsat-4 would approximate the 80 m x 80 m ground area of Landsat-1, -2, and -3 pixels. Landsat-5 will be identical in this regard to Landsat-4.

6.2.5. *Data Handling*. Since 1972, when the first Landsat satellite was launched, the data handling part of the system has been constantly evolving and will continue to do so now that Landsat-5 is functioning successfully. Essentially, data acquired by the MSS system on Landsats-4 and -5 are transmitted either on a real-time basis to one of NASA's three ground receiving stations. Equipment at these stations arrange the data and record them on videotape. Each morning the data received at the Fairbanks and Goldstone receiving stations are transmitted to NASA/Goddard via Domsat, a commercial domestic communications satellite system.

Processing of Landsat data consists primarily of converting the video data to a digital format and recording them on high density tape. The tapes are then processed through the Master Data Processor (MDP) which performs some radiometric and geometric corrections. The partially corrected data are transmitted to the EROS Data Center (EDC) in Sioux Falls, South Dakota, via

Domsat. Landsats-4 and -5 MSS data are produced at NASA/Goddard by the National Oceanic and Atmospheric Administration (NOAA).

Data received at the EDC are recorded on high density tape. The tapes are converted by the EROS Digital Image Processing System (EDIPS) into photographic and digital products. A laser-beam film recorder is used to produce film negatives. The EDIPS system is capable of resampling and applying geometric corrections to Landsat data. This allows the center to establish an archive to radiometrically-corrected data only for customers who prefer working with 'raw data'. These data are provided as computer compatible tapes (CCT's).

Data handling for Landsats-4 and -5 are still in the evolutionary stage. It is anticipated that when the system is completed, digitized data from the MSS and TM will be transmitted on a real-time basis to one of two Tracking and Data Relay Satellites (TDRS) positioned in geostationary orbits over the equator, and from there to a ground-receiving station at White Sands, New Mexico. Data will be relayed from White Sands to NASA/Goddard via the Domsat system. From Goddard, corrected MSS data will be transmitted to the EDC; no transmission of TM data between Goddard and the EDC is planned. TM products, including both 240 mm film and computer compatible tapes, will be produced at Goddard and shipped via air freight to the EDC for distribution. The first TDRS vehicle (TDRS-A) is now in a parking orbit over the Atlantic, just off the east coast of Brazil. A second (TDRS-B) will be launched in late 1984 and placed in orbit over the western Pacific Ocean.

In addition to the CCTs Landsat digital data, or at least those acquired since February 1979, may be obtained on floppy disks for use with microcomputers. Each disk contains a user-specified subscene extracted from a standard fully-corrected Landsat MSS digital tape. The ground area of the subscene must fall within a 240- by 256-pixel matrix. All four spectral bands of the subscene are provided in a band-interweaved-by-pixel format. The disks are 8-inch, CPM compatible. When ordering the user must identify the desired subscene by means of a transparent grid that can be obtained from EROS Data Center for one dollar. The cost of a typical disk, consisting of a 240- by 256-pixel area in all four bands, is presently one hundred dollars. Additional disks from the same scene cost considerably less.

From the user's viewpoint, then, there are three types of Landsat products available: imagery, CCTS, and floppy disks. Which of these is appropriate to any task will depend upon the data needed and the time and money available. Questions about Landsat products should be directed to:

NOAA Landsat Customer Services
Mundt Federal Building
Sioux Falls, south Dakota 57198
Telephone: (605) 594-6151

6.3 Image Interpretation

Landsat data are digital, not photographic. Nevertheless, prints made from Landsat data resemble aerial photos because the digital data have been converted to colors and/or gray tones at such small scales that the individual pixels cannot be discerned. One result is that Landsat imagery may be analyzed using the same aerial photointerpretation techniques and equipment. This approach was commonly used in the early days of the Landsat Program before the development of sophisticated software packages made computer analysis of Landsat data so efficient. Nevertheless, there remain a variety of uses for the imagery: compiling thematic maps, updating existing maps, making certain kinds of measurements, and presenting graphical information. They may often substitute for mosaics. In general, Landsat imagery acts as a complement to aerial photography.

From the beginning the most frequently used product has been the standard 18.5 cm (7.3 inch) MSS image, scale 1 : 1 million (Table 9). For each scene, five such images are available – four black-and-white and one false-color composite (FCC). The four black-and-white images represent the four spectral bands – green, red, and two reflective infrared. The fifth, the FCC is a simulated color-infrared image produced by the superimposition of the green, red, and one reflective-infrared band (usually band 7). With the exception of the FCC all MSS products are available as paper prints, film positives, and film negatives. The false-color composite is available only as a paper print or film negative. In addition to the 18.5 cm format, images may be ordered in a 37.1 cm (14.6 inch) or a 74.2 cm (29.2 inch) format. These larger products, though considerably more expensive, are also more useful because of their larger scale.

While the typical Landsat MSS image may resemble a high-altitude aerial photo, there are some rather obvious differences between them. Figure 40 depicts the coverage of a MSS image. Notice first the shape of the image; it is not a square but rather a parallelogram. This shape is due to the fact that it took the scanner 25 seconds to acquire the data presented on the image. During that interval the earth was rotating beneath the spacecraft. Thus, the image is skewed

Table 9. Prices of standard Landsat MSS imagery*

Nominal image size	Scale	Product	Unit price B&W	Unit price Color
18.5 cm (7.3")	1:1 million	Paper	$30	$45
18.5 cm (7.3")	1:1 million	Film positive	$30	$74
18.5 cm (7.3")	1:1 million	Film negative	$35	N/A
37.1 cm (14.6")	1:500,000	Paper	$58	$90
74.2 cm (29.2")	1:250,000	Paper	$95	$175

*As of 1 May, 1984.

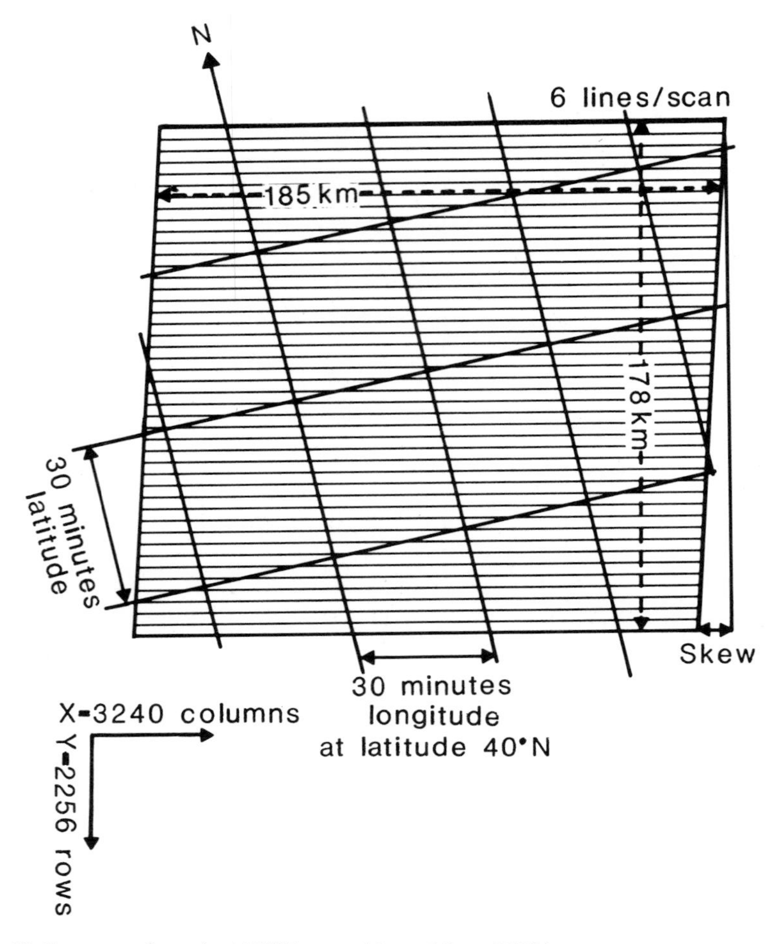

Fig. 40. Coverage of standard MSS image. Adapted from NASA.

by a distance equal to the earth's rotation in a 25 second period. But the shape is not the only thing skewed; so, too, are the lines of latitude and longitude. This results from the satellite's orbital path which is inclined 99° from the equator.

Scale and resolution provide two additional noticeable differences. At a scale of 1:1,000,000 a single Landsat image covers an area of 34,000 square kilometers (13,000 square miles), an area no aerial photo can begin to encompass. In fact, at 1:15,000 scale (6-inch focal length camera) with the usual 60 percent endlap and 30 percent sidelap, it would require 5,000 photos to cover such an area. This large area perspective, then, is one of the qualities of Landsat that makes it so valuable. Many features and relationships not apparent on large-to-medium scale aerial photos become visible on Landsat imagery. Contributing further to this perspective is resolution. Whereas on a typical aerial photo ground

Fig. 41. MSS band 5 (red) image of southern New England. Scale 1 : 1 million. Courtesy NASA.

or spatial resolution may be measured in a few meters (or feet), on MSS imagery it approximates 80 m (perhaps 40 m for certain linear features). The effect of such resolution is to eliminate the meticulous detail of aerial photos and replace it with a more vivid image of the larger, and often less visible features.

Figure 41 depicts an MSS band 5 (red) image of southern New England. Along the outside image edge, latitude and longitude tick marks are depicted at 30-minute intervals. The reference marks are annotated in degrees, minutes, and compass direction. At the bottom a 15-step gray-scale table appears, as it does on every frame of Landsat imagery. The scale is used to monitor and control printing and processing functions and to provide a reference for analysis related to a particular image. The annotation block directly over the gray scale, provides the information relevant to image identification, geographic location of scene, and time of acquisition. Reading from left to right the annotation may be explained as follows:

10 OCT 72 (calendar date of Landsat frame data acquisition)

C N40-18/W074-51 (latitude/longitude coordinates of scene principal point)

N N40-17/W074-48 (latitude/longitude coordinates of nadir point)

MSS 5 (sensor and band number)

D (represents direct transmission as opposed to R stored on WBVTR and played back)

SON EL 38 AZ 150 (elevation and azimuth of sun at time of data acquisition)

191-110-N (spacecraft heading; orbital path)

I-N-D-2L (indicates various processing procedures)

NASA ERTS (Agency and project)

E-1079-15131-801 (frame identification numbers)

Like many aspects of the Landsat Program the annotation block has been continuously evolving, and data acquired after 1978 will be annotated slightly differently.

As to the image itself, band 5 is generally the best for emphasizing cultural features. RBV imagery from Landsat-3 is available at the same price as MSS imagery. RBV imagery, while not as numerous, has some distinct advantages over MSS. Resolution is one; RBV imagery from Landsat-3 has a resolution of approximately 30 m. As pointed out previously, one RBV scene covers an area one-quarter that of an MSS. Furthermore, because the RBV system images a scene simultaneously, it captures an undistorted view of the earth's surface (except for the distortion caused by the earth's spherical shape). With the termination of Landsat-3 the RBV system also terminates.

6.3.2. *Digital Data Analysis*. Digital data analysis is a complex and costly process, particularly when compared with conventional photointerpretation techniques including Landsat image analysis. Digital analysis requires either computer training or access to a computer programmer or image analyst; it also requires access to a computer. The cost of purchasing a computer for this purpose could run to hundreds of thousands of dollars depending upon type, capacity, and various input and output devices, and would hardly seem worth it except for government agencies and large private firms. A better option might be to employ an existing computer, possibly one owned by a state agency or university, or to purchase computer processing services from a private firm. In any of these cases, the appropriate software must be acquired. Information about available software packages may be obtained from COSMIC – Computer Software Management and Information Center, University of Georgia, Athens, Georgia.

While the cost of digital analysis is admittedly great, at least in terms of Landsat data, it is the most effective approach. Its advantages include the ability a) to manipulate large data sets; b) to store and retrieve these data instantly; c) to precisely measure small differences in spectral values; d) to duplicate results; and e) to integrate these data with other non-remote sensing-derived data. These qualities suggest, however, that the digital approach is best for large area analysis.

The computer tapes (CCT's) necessary for digital analysis are generated on request from high density tapes maintained at NASA/Goddard. The standard CCT's are today 9-track magnetic tapes on which data are packed either at 1600 bits per inch (BPI) or 6250 BPI. At 1600 BPI the four spectral bands representing one Landsat scene (30 million pixels of data) are written on one tape. And with the launch of Landsat-3 a new 6250 BPI tape has been introduced which is capable of handling up to 100 Landsat scenes.

The processing of Landsat CCT's may be undertaken in one of two ways – batch mode or an interactive mode which utilizes some kind of graphics system such as a cathode ray tube (CRT). Batch processing involves submitting a project to a central computer and waiting for the results. Since the computer is handling several jobs simultaneously, the output may not be returned until the next day. Batch processing can be very slow, particularly if the user must depend on a programmer for assistance. On the other hand, compared to the interactive processing mode it may also be considerably less expensive.

The interactive approach allows the user more direct involvement in the processing operation and for this reason is preferred to the batch mode. After inputting data and the proper commands, results are displayed on the screen almost immediately; no prolonged wait is necessary. In addition, the user may quickly make modifications in the data or commands; portions of the image may even be enlarged for closer scrutiny. There are a number of commercial interactive systems on the market, including G.E.'s Image 100 and Bendix Corporation's MDAS. The major disadvantage to all these systems is cost; prices generally run into the hundreds of thousands of dollars.

Regardless of the approach or system, computer analysis can be broken down into three categories or tasks: 1) preprocessing, 2) enhancement, and 3) classification. Each of these tasks is complex and the discussion here will do little more than touch on them. Therefore, for a more thorough explanation, Nicholas Short's *The Landsat Tutorial Handbook: Basics of Satellite Remote Sensing* (1982) is highly recommended. The book, NASA Reference Publication 1078, may be obtained from the United States Government Printing Office.

The first step in the computer processing of Landsat data is referred to as preprocessing. It involves a series of operations to eliminate distortions introduced by the atmosphere and the system itself. For example, radiometric corrections are made because the MSS system's 24 detectors do not respond identically. The effect is to cause uneven lines or stripes to appear on MSS imagery. While these variations do not significantly affect conventional photointerpretation of the imagery, they may well influence the digital analysis. Geometric corrections are necessary if data are to be transferred to map bases or geobased information systems, and there are many to be made. Geometric distortions are introduced by such factors as the curvature and rotation of the earth, variations in the altitude and velocity of the spacecraft, and delays in detector sampling. Additional corrections must be made for atmospheric haze that may actually vary from scene to scene. Here the effect is to increase radiation values, particularly in the shorter wavelengths. Haze removal algorithms subtract an appropriate brightness value from each spectral band in order to make the data more comparable. Preprocessing is routinely done on Landsat data both at NASA/Goddard and the EDC. On occasion, however, a user may request specifically that tapes be unprocessed in order to analyze the unaltered data.

The purpose of operations generally included under the heading 'enhancement' is to provide Landsat imagery for analysis by conventional photointerpretation techniques. Among the routines applied are density slicing, contrast stretching, and ratioing. Density slicing involves the combining of several different values into a single value. This can be an effective method of enhancing a particular type of land use, for example, if that land use has a unique and rather narrow range of values. The new value is assigned a specific gray-scale value or color, while those values not falling within the range of the compressed values are assigned another. It is possible to individually map land uses in this way.

Contrast stretching, by comparison with density slicing, involves increasing the range of brightness values, though for the same purpose. Since incoming radiation is usually concentrated in a narrow portion of the total range over which the system can receive it, computer routines are used to increase some values and decrease others. The effect is a sharper, more pleasing image. Both density slicing and contrast stretching work best on single-band scenes.

Ratioing involves the use of two spectral bands. Reflectance (brightness) values of one band are divided by the reflectance values of another to produce a quotient for each pixel. This is done most commonly between bands 4/5, 5/6, and 6/7. Black-and-white images can be produced from each of these ratios. This is accomplished by assigning a specific gray-scale value to each quotient. It is also possible to produce a color composite by using three sets of ratio images. The overall effect of ratioing is to smooth out tonal contrasts resulting from topographic differences.

Most applications of the computer processing of Landsat data require that the various features comprising the earth's surface be classified into meaningful categories. This is accomplished by a process referred to as spectral pattern recognition, although it is actually more 'spectral' than 'pattern' recognition; in fact, most classification methods rely very heavily on spectral signatures for feature identification. Unfortunately, many features and land uses do not have unique spectral signatures but rather have a wide range of values. Hence, the classification process is complex. Complicating it further is the frequent presence of several features and land uses within one pixel, a common occurrence in urbanized areas. The combined reflectance values from spectrally different surfaces may make it difficult to assign such a pixel to any specific category;

even worse, it may fall within the spectral range of a very different feature or land use type.

While there are a host of techniques for classifying spectral data into statistically separable classes, most utilize one of two classification methods. These are known as the unsupervised, or clustering, method and the supervised, or training, method. Each has its advantages and disadvantages.

The unsupervised method involves the use of clustering algorithms to automatically sort data into spectrally similar classes (clusters). Every pixel is assigned to some class on the basis of distance from the cluster center. The clusters represent nothing more than ranges of pixel values until they have been related to terrain features either through field observation or by comparison to aerial photos. When this is done, it is often found that a given feature is represented by more than one cluster or, more commonly, that more than one feature or land use is described by one cluster. The process may have to be repeated to resolve these problems. This method is utilized most frequently when little ground truth is available and the mix of surface features is great. It is also one that can be performed by private firms offering Landsat services.

The supervised method classifies data according to spectral signatures that have been predetermined. This is accomplished by establishing 'training' sites for the computer. By examining the spectral signatures for the dominant features of the training sites, ranges of values are generated for each type of feature or land use. These values are gradually extended to the remainder of the scene until all data points have been classified. Inevitably, many data points or pixels will be misclassified, so it is customary to periodically view the output on a CRT or printout. Categories are checked against 'ground truth' data, usually aerial photos, and, if necessary, the range of spectral values for any particular feature or land use class may be modified. This process continues until errors are reduced to an acceptable minimum at which time a hard copy of the output is made. The usual output is a land cover/land use map generated either in black-and-white on a line-printer, or in color using a color film recorder. Though this is an expensive method, the continuous interaction between user and computer combined with a rapid turn-around time cannot be matched by the unsupervised approach.

In summary, there are a variety of advantages to the Landsat system that make it attractive to land planners. Though supporting data are not abundant, it appears that for certain tasks Landsat may represent a cost-effective alternative to conventional ground and aerial surveys. One such task is land cover/land use mapping over large areas. The Southwestern Illinois Metropolitan Regional Planning Commission reported that a land cover inventory using Landsat computer tapes cost a fraction of what it did to conduct the same inventory from aerial photos and an automobile windshield survey (Table 10) though costs of Landsat products have increased dramatically since the study was completed. A

Table 10. Cost comparisons for land cover inventories*

	Ground survey**	Aerial photo-interpretation***	Landsat****
Maps, photos or CCT	$ 130.90	$ 4,161.38	$ 400.00
Auto travel expense	570.90	Minimal	700.00
Inventory	55,000.00	17,056.00	3,792.00
Measure and tabulate	30,000.00	8,528.00	3,700.00
Ground checking			750.00
Map preparation	20,000.00	5,969.00	5,615.00
Miscellaneous supplies	500.00	135.00	1,000.00
Total	$106,201.80	$36,049.38	$15,957.00
Per square mile	$ 59.46	$ 20.18	$ 4.20

**A Legislator's Guide to Landsat*, Denver: The National Conference of State Legislatures, 1979, p. 5.

**Eight years, 40 categories, 3 counties, 1786 square miles.

***Eighteen months, 5 categories, 3 counties, 1786 square miles.

****Six months, 16 categories, 7 countries, 3792 square miles.

second advantage is the capability of Landsat to acquire data continuously in four or more spectral bands. Because of their comparability, parallels can be drawn between data from different spectral bands, and between data acquired from the same spectral bands but on different dates. Finally, Landsat is capable of acquiring data over large areas and in a form convenient for computer analysis. Landsat can reveal where more information is needed. In those areas where land parcels may be small and the mix of cultural features great, Landsat will give way to aerial photos and ground surveys. Thus, Landsat complements other data sources at both the acquisition and analysis stages.

6.4 How To Order Landsat Products

Landsat products are available for the fifty states and for most of the earth's surface outside the United States. They can be ordered through the U.S. Geological survey's National Cartographic Information Center (NCIC) or the EROS Data Center (EDC). When ordering, it is necessary to supply either the geographic coordinates or a map marked with the area of interest. Use of the U.S. Geological Survey's Geographic Computer Search Inquiry form is encouraged. In addition, it is necessary to specify the type of Landsat product (18.5 cm black-and-white film positive), the minimum image quality acceptable, the maximum percent of cloud cover acceptable, and the preferred time of year. For ordering imagery of the United States, the form 'Selected Landsat Coverage' may be helpful because it provides a map illustrating the path and row of individual scenes.

It is important to be aware of the fact that all orders must be prepaid or made on a company or government purchase order. Additional information may be obtained from the following centers:

National Cartographic Information Center
U.S. Geological Survey
507 National Center
Reston, Virginia 22092
Telephone: (703) 860-6045
or
National Earth Satellite
EROS Data Center
User Services Section
Sioux Falls, South Dakota 57198
Telephone: (605) 594-6511

6.5 Postscript

With the launch of Landsat-4, responsibility for the management of the Landsat Program has been transferred to the National Oceanic and Atmospheric Administration (NOAA) since by charter NASA cannot maintain an operational system. NOAA has managed the nation's operational weather monitoring satellite programs (METSAT) since their inception in the early 1960's. As manager of the Landsat Program, NOAA will be responsible for spacecraft scheduling, ground processing and production, and customer services. All requests for Landsat data will continue to be forwarded to the EDC where they will be separated into retrospective requests and special acquisitions. Retrospective requests will be filled directly from archival materials; special acquisitions will be forwarded to spacecraft command at NASA/Goddard for spacecraft scheduling.

The longer term future of Landsat is less certain. The Reagan administration has been hoping to sell the Landsat system to the private sector as an economy move. Accordingly, funding for future satellites has been slashed and only the launch of Landsat-D' (Landsat-6) is now assured. While recent Congressional criticism to the proposed sale has forced its postponement, NOAA's funds for the Program have not been replaced. Furthermore, since NOAA was forced to launch Landsat-5 ahead of schedule, it faces the possibility of a gap in data acquisition around 1985 or 1986.

With this possibility in mind some Landsat users are looking to France's Systeme Probatoire d'Observation de la Terre (SPOT) satellite as a future alternative. Expected to be launched in 1985, SPOT will incorporate several technical innovations that will enhance its value for mapping and monitoring land resources [18].

For one, the SPOT system will consist of two identical high resolution visible (HRV) imaging instruments. Representing the latest in solid-state scanning

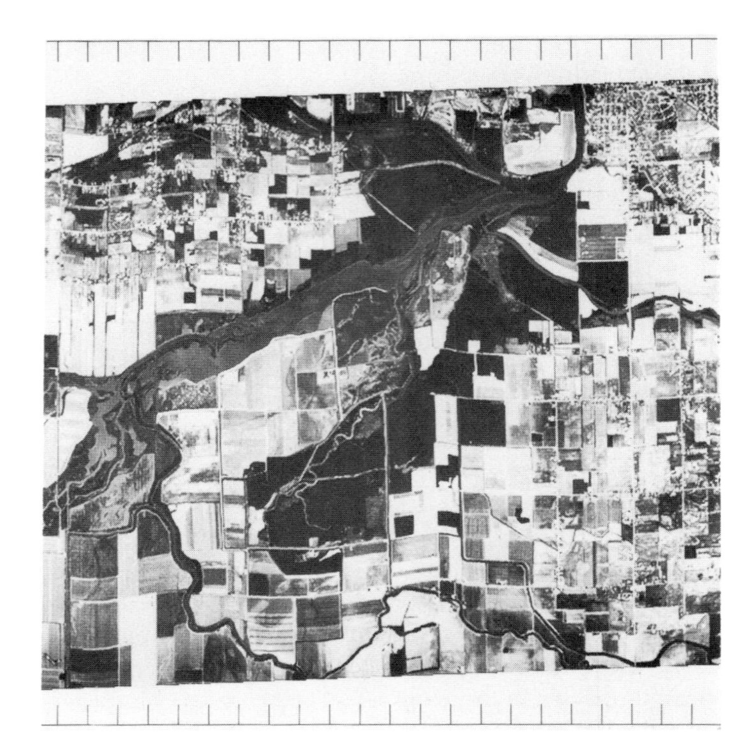

Fig. 42. SPOT data simulation. Simulated SPOT image of Saginaw County, Michigan. The image, a band 2 (red), has a ground resolution of 20 meters. Copyright 1983 SPOT Image Corporation.

devices, each instrument contains 6000 detectors arranged in a linear array. Although known as 'pushbroom' scanners, the HRVs do not scan but acquire a complete line of ground data in a single look. For this reason they are expected to provide more reliable and accurate data than Landsat's MSS.

The HRVs will also provide higher resolution data than Landsat. When operated in a panchromatic mode they will provide 10 meter resolution data, when operated in a multispectral mode – 20 meter resolution data (Fig. 42).

Finally, and perhaps most interestingly, the HRVs will be steerable, that is, by using tiltable mirrors adjustable on command, they will be capable of reimaging an area any time within 4–5 days of the initial pass. The SPOT system will therefore have the capability of acquiring stereo imagery. With modification Landsat ground stations will be able to receive SPOT transmissions.

7. Remote sensing input to geographic information systems

7.1 Introduction

Land planning in the United States is conducted by a bewildering array of federal, state and local officials, with the bulk of the responsibility resting at the local level. This fragmentation of responsibility has led to a multitude of approaches to the planning process with little attendant agreement on what the planning process should even involve. For the purposes of this book, the planning process will be viewed as ideally consisting of several steps. These include:

1. the establishment of long-term goals for the use of land resources;
2. the gathering of information about those resources;
3. the evaluation of alternative strategies for meeting long-term goals;
4. the designing of policies to meet these goals;
5. the adoption and implementation of these policies;
6. the monitoring and continuous reevaluation of these policies as a means of attaining the long-term goals.

Assuming some agreement with these steps, remote sensing can be seen as making an input to the planning process at several points. First, remote sensing may provide a cost-effective method for acquiring many of the data necessary for the establishment of land resource policies. Information on land capability, land use, wildlife habitats, new subdivisions, and transportation accessibility are just a few of the things that can be obtained quickly and efficiently by remote sensing. At the adoption stage, enlarged and annotated aerial photos may be used at planning board meetings and public hearings to more effectively explain the policies being considered. And finally, at the reevaluation state, aerial photos may be used to gather information at regular intervals as a means of monitoring the effects of these policies.

For information (remote-sensing derived or other) to be useful to policy makers, it must be properly managed. Information management involves the acquisition, storage, retrieval, manipulation, and distribution of data to those who need them. A geographic information system is one in which the data have been georeferenced. Geographic information systems are basically of two types, manual or automated.

7.2 Manual Geographic Information Systems

The primary method for storing, manipulating and presenting data in a manually-accessed geographic information system is the base map. Base maps are available in a variety of formats. Photogrammetric line maps provide the most common data bases, but rectified enlargements of aerial photos, orthophotos, and topographic maps are also used.

Photogrammetric line maps (see Fig. 43) are produced by aerial survey and engineering firms from precision aerial photography, ground control data, and stereoplotting equipment. Theoretically, any feature discernible on an aerial photo may be plotted, including contour data. In practice, only street lines, major rivers and streams, and property lines are usually displayed. The final map is compiled in ink on stable base mylar which allows for easy reproduction by an ozalid printing process. Photogrammetric line maps meet National Map Accuracy Standards and are generally provided at large scales (1 : 1500 and larger). Contour data are typically represented at 1- or 2-foot intervals.

Aerial photo bases, either in the form of rectified enlargements or orthophotos, are not only becoming more common today but are recommended as preferable to the line map. Data are compiled on overlays to the photo base. There are a number of advantages to the use of the photo base/overlay approach. For one thing, data appearing on the overlays can be seen in their relation to landscape features. For example, property lines can be seen to follow stone walls, fence lines, and streams. Furthermore, if property lines are being determined from the original deeds, not only are landscape features visible but on orthophotos distances and directions are true. Another advantage to the photo/overlay approach is the ease in keeping information up-to-date. Periodic reflying of the photo base should reveal changes in land use such as new housing, land abandonment, and new highway construction. These changes can be added to existing overlays or to new overlays as a means of monitoring change.

Aerial photo bases may be provided at literally any scale. It is often useful to produce them at several scales in order to meet the demands of various users. Suggested scales might be: a large scale (1 : 500 or 1 : 1000), an intermediate scale (1 : 5000), and a small scale (1 : 10,000). Photo bases are usually provided in a 30 in. × 30 in. format with overlays made of reproducible stable base mylar. Individual overlays can be made for contour data, soils, drainage, utility lines, land use – whatever is deemed important.

USGS topographic maps make useful bases when smaller scales are required. The 1 : 24,000 ($7\frac{1}{2}$-minute) quads are available either as orthophoto maps or conventional line-and-symbol maps. A 1 : 50,000 is available which may be formatted by county or some other administrative unit rather than by individual sheets. Finally, there is a relatively new 1 : 100,000 series available from the USGS's

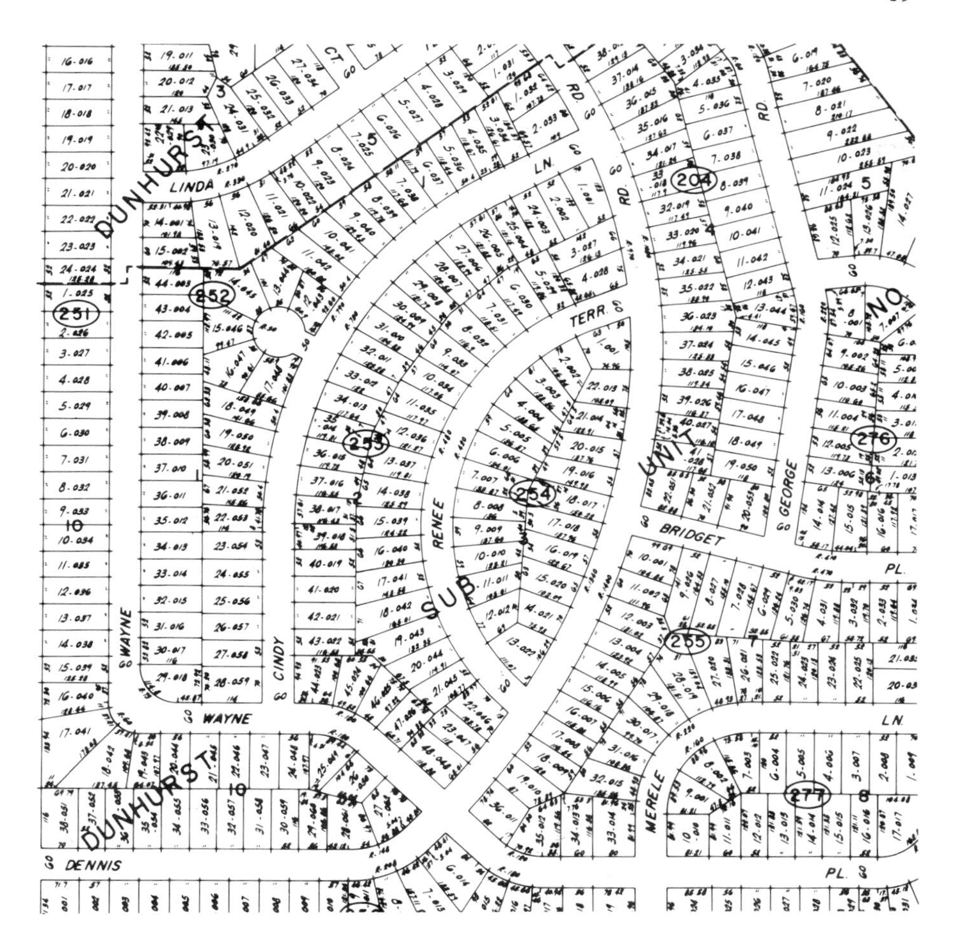

Fig. 43. Line map showing property boundaries, dimensions of each parcel, and parcel identification number. Courtesy Sidwell Company.

National Mapping Division. All three have the advantage that the necessary locational references information is already provided on the maps.

7.2.2. *Case Study*. A good example of the multi-scale data base approach is presented by Vermont where, in 1967, a program was instituted to provide each city and town in the state with its own base maps. Ostensibly the maps were to be used for tax assessment purposes and as such would show land ownership, dimensions, and acreage. It was anticipated that the maps would provide many communities with the basis for a manual information system that could be used for land appraisal, planning, and resource management.

The Vermont Mapping Program, as it is officially known, involves the production of maps at four scales: 1:5000, 1:2500, 1:1250, and 1:625. The particular sequence of scales was chosen for reasons of simplicity. Since it was anticipated that there would be frequent need to move back and forth between scales, it was decided to reduce the chance of human error. Linear dimensions need only be doubled or halved while for aerial entities the relationship is either times- four or one-quarter. The same rationale led to not using metric scales.

Two of the scale series, the 1:5000 and the 1:2500, are orthophoto maps. The 1:2500-scale sheets are enlarged by precision camera from quandrants of the 1:5000-scale sheets and are then reformatted. The other two scales, 1:1250 and 1:625, are photogrammetric line-and-symbol maps. The 1:625-scale sheets are made by camera enlargement of quandrants of the 1:1250-scale sheets, redrafted and reformatted to the same standards as the 1:1250-scale sheets. The 1:5000-scale orthophoto maps and the 1:1250-scale line-and-symbol maps are constructed to meet National Map Accuracy Standards.

The 1:5000-scale maps cover the entire state. Where areas contain too many properties to effectively delineate them to the 1:5000 scale, the 1:2500-scale sheets are provided. The 1:1250 line-and-symbol maps displaying buildings, roads, railroads, utility lines, and the like are available only for the most densely settled areas (cities and larger towns). Where the latter areas contain parcels too small to be properly mapped at the 1:1250 scale, 1:625-scale sheets are used.

All maps are laid out in metric intervals of the Vermont Coordinate System. In addition, grid ticks of other map projections appear at the edges of the photos. These include the geographical coordinate system with markings for degrees, minutes, and seconds of latitude and longitude and the Universal Transverse Mercator System. The latter is marked in increments of 500 meters.

To facilitate their use, a multi-unit bar scale appears on all maps. This bar scale consists of units showing (at the scale of the map) meters, kilometers, feet, yards, rods, chains, miles fractional, and miles decimal. Therefore, whatever unit of measurement is being worked from deeds, plans, plats, or surveys, the user experiences, at least in theory, no difficulty as long as the proper unit of the bar scale is used.

A final but vital element of the Vermont Mapping Program is map maintenance and updating. It is anticipated that every few years orthophoto overlays will be produced under contract for those areas of rapid development. The overlays will remain a significant part of the system as a source for verifying changes in land use.

Additional information on the Vermont Mapping Program may be obtained by writing: Director, Division of Property Valuation and Review, 43 Randall Street, Waterbury, VT 05676.

7.3 Automated Geographic Information Systems

For agencies of local governments such as planning boards, zoning boards, and conservation commissions, manual information systems are generally adequate.

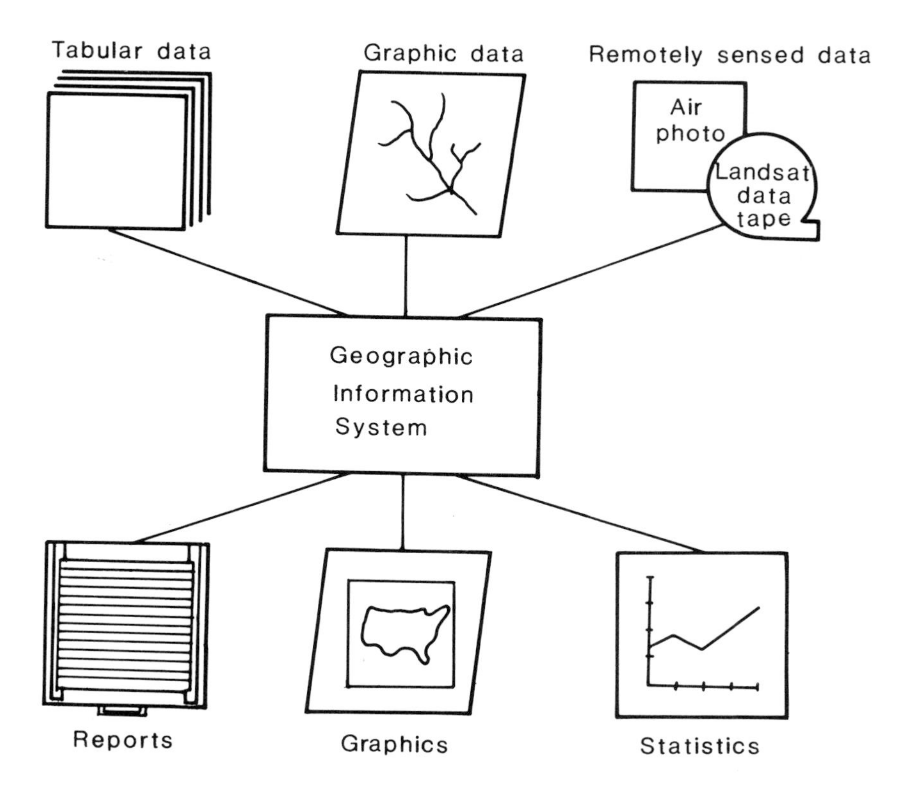

Fig. 44. Schematic diagram of a typical Geographic Information System. Adapted from Short (1982).

They are also often the only affordable option. For municipalities and states, manual information systems are not adequate for effective and timely decision-making. The sheer quantity of information to be managed requires the use of computerized data-handling techniques. Accordingly, a growing number of municipal and state governments have established automated geographic information systems for land planning and resource development purposes.

As illustrated in Figure 44, an automated geographic information system is basically a computerized filing system into which data are entered, stored, processed, and presented – all in a format appropriate to the user's needs. The data input to a system may originate from a variety of sources, for example, historical documents, property deeds, conventional maps, field surveys including those conducted by federal agencies such as the Bureau of Census, and remote sensing. The data may be tabular, graphical, or image. Whatever their source or format, they must be properly encoded before being input to the system.

There are two widely used formats for encoding and storing land data – the polygon (vector) and the grid cell (raster). Polygon-based systems are often used for small-area planning where the major concern is for the accurate cartographic retrieval of data and for area measurement. The grid system tends to be preferred for large-area planning, where accuracy is not as important as being able to overlay and compare several data sets. An ideal system would be one that could display both and covert from one to another [19].

Polygon systems depict geographic areas (i.e., census tracts, political units, land use types) as closed chains of straight line segments. The areas are transformed by determining the vectors (x, y coordinates) of the nodes or pivot points connecting the straight line segments. Linear features, such as rivers, are recorded in a similar manner. Thus landscape features are encoded as sequential lists of vector data. While the system is accurate in defining the shapes of areas, it is expensive due to the complexity of data encoding and search procedures, as becomes evident when two or more data sets are compared.

The purpose of the grid system is to transform information found on maps and aerial photos into a machine-readable format. It works particularly well for describing phenomena that occur continuously over large areas such as soil types and land uses, though not so well for political boundaries. Data are encoded by placing a transparent grid over the area of interest and recording the dominant class of information for each cell on a line-by-line or raster basis. The smaller the size of the grid-cell, the finer the detail extracted from the source material. If the cell size is enlarged, the accuracy of the description of the original data will lessen accordingly, though this may be offset somewhat by the need for less data storage. Cell sizes of under 10 acres are usually required for most local planning purposes. At the state level, information systems typically employ cell sizes of 40 acres or more.

Data stored in raster format may be merged to perform a variety of multi-variable analyses. The data need not be stored on the same magnetic tape. It is necessary only that the several data sets be registered to the same grid. Computer software will cause the required data sets to be superimposed one upon the other in a manner similar to the manual overlaying of transparent maps. Since each data set represents a designated layer, each variable can be accessed by specifying the appropriate layer. Specifying the row and column can locate a specific grid-cell (or land parcel) within the designated layer. A common application of this type of analysis is the land suitability/capability study (Fig. 45). The user may request that for a certain area data on geology, slope, soils, erosion potential, land use, and anything else deemed important be overlayed with weights assigned to each variable. The computer will then produce a map of land suitability according to the constraints posed by the user. The weights can be changed and another map produced to determine what effect the weights have upon the map.

Since it is far simpler to overlay raster data than vector data, the latter are often converted to raster format for this purpose. Political boundaries, census tracts, and traffic zones are typically input as vector data. Their conversion to

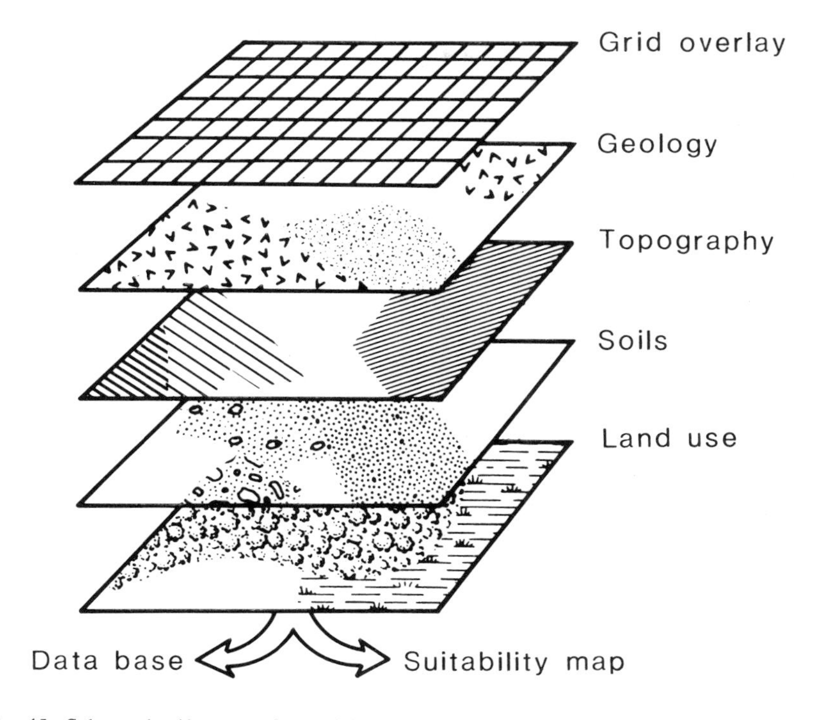

Fig. 45. Schematic diagram of a multiple variable data base.

raster format allows for a variety of numerical and geographical analyses. If land use data are on-line, for example, they can be reported out in acreages or percentages by county, township, or census tract. They may also be merged for comparison with other socioeconomic data, or be displayed graphically.

To accomplish these things, geographic data must be referenced to a specific location on the earth's surface. This is usually done in one of two ways – using ordinal codes as indices to location or using nominal codes. With ordinal codes, locations are typically expressed as coordinates on some type of geographic reference system. The most commonly employed ordinal systems are latitude and longitude, the Universal Transverse Mercator (UTM) grid, and the State Plane Coordinate System (SPC). Nominal codes do not use coordinates but 'names', as the term implies. Of the various nominal codes census tract numbers, street addresses, and ZIP codes are the identifiers most often used. Since nominal codes do not give exact locations of data within cells, some geographic accuracy is lost.

Until recently, the development of automated geographic information systems was limited to those federal and state agencies having access to large and

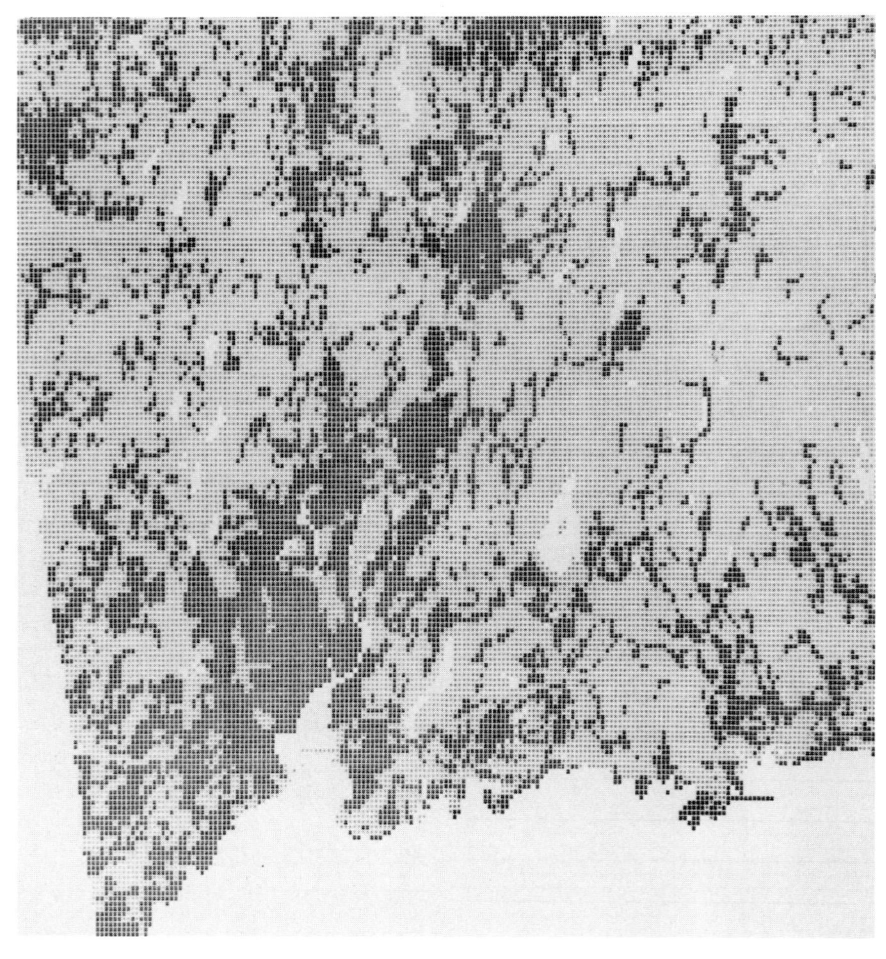

Fig. 46. Lineprinter map illustrating the built-up area of metropolitan New Haven, Connecticut.

expensive mainframe computers. But now the situation has changed. Less expensive minicomputers are on the market that possess many of the same capabilities as mainframes, including large data storage. Thus, most state and many regional and municipal planning agencies have developed, or are in the process of developing, their own automated geographic information systems. And with the advent of even less expensive microcomputers like the Apple II Plus and the IBM Personal Computer local planners too have the option of developing such systems. Although by definition microcomputers are small and limited in their computational capabilities, they are able to handle most of the tasks required of local planning officials. The case for microcomputers is further strengthened by the fact that Landsat digital products are now available on floppy disk. Each disk will contain data of a ground area approximating that covered by a $7\frac{1}{2}$-minute quad sheet.

There are a variety of outputs that can be generated from automated information systems. The simplest output is the printout from a teletype or line-printer which portrays geographic information through the use of alphanumeric symbols and/or overprinting (Fig. 46). SYMAP, developed by Harvard University's Laboratory for Computer Graphics, is the best known system of this type. Although maps produced in this manner are relatively inexpensive, they contain distortions in scale and are often difficult to decipher. Occasionally, it is necessary to hand-color them before certain geographic patterns can be recognized. Teletypes and lineprinters are also limited in the number of gray tones they can produce; ten to twelve appears to be the maximum. As a result of these various limitations, teletype and lineprinter maps are best used for preliminary types of analysis or for previewing final maps before producing them in a more expensive format.

There are numerous peripherals available for generating output from automated geographic information systems. Plotters have the capability of producing high resolution maps from vector data. One type, the drum plotter, does so by means of a stationary plotting pen and a rotating drum onto which drafting paper has been affixed. The drums are typically 11 or 30 inches wide and can handle paper of infinite length. Although older versions of drum plotters produce maps displaying 'steplike' lines, newer models produce smoother lines without visible incrementing [20]. A second type of plotter, the flatbed, produces maps by the movement of a drawinghead across paper affixed to a horizontal table-top. Both types allow lines either to be drawn in liquid ink or ballpoint, scribed on a scribing film, or exposed on photographic film. While more expensive to produce than lineprinter maps, plotter-derived maps have more precise line work and are therefore easier to read and aesthetically more pleasing (Fig. 47).

Increasingly, today, geographic information systems are being connected to graphics display terminals or CRT's which allow data to be quickly displayed and easily edited. Unfortunately, these display screens are less suited to cartographic display than are alphanumeric systems, because they lack fine resolution. The technologies in this field are changing every day, however. Another drawback to display screens has been the difficulty in providing hard copy for publication and analysis. The image appearing on the screen will be lost unless captured in some way. Some display terminals have been interfaced with electrostatic plotters to produce an output, in either black-and-white or color, similar to that provided by an office copier. Additional hard copy devices include instamatic cameras which may provide output as prints, transparencies, or 35-mm slides, and microfilm recorders, which employ a laser beam technique to expose fine lines on 35-mm film.

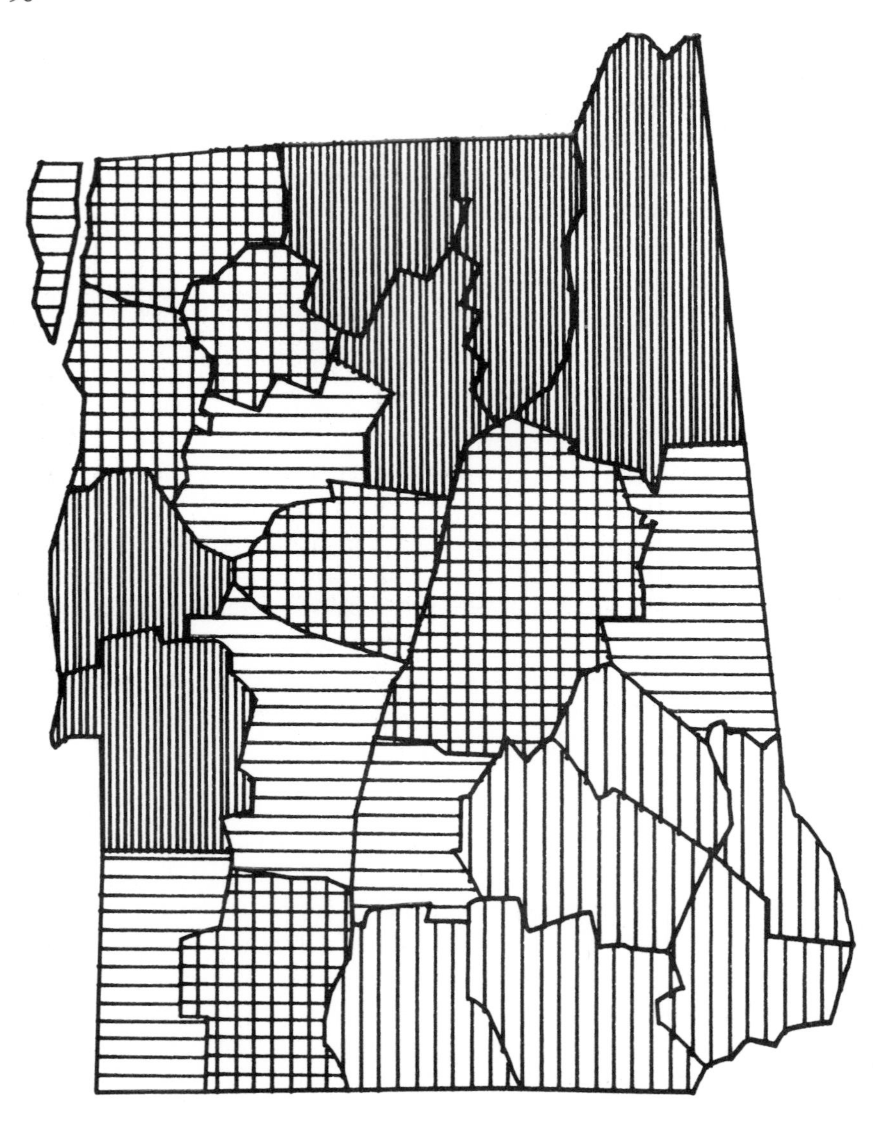

Fig. 47. Map from flatbed plotter illustrating mean farm size per county for states of Vermont and New Hampshire. Narrowly spaced vertical lines represent farms of 190–220 acres, gridded areas farms of 165–189 acres, horizontal lines farms of 132–164 acres, and widely spaced lines farms of 83–131 acres. Courtesy David Jensen.

Ultimately, the effectiveness of a geographic information system rests upon the capabilities of the software. Ideally the software should enable the user to: 1) retrieve one or more data elements from a file; 2) transform or manipulate the values in any data element retrieved; 3) transform, manipulate, or combine all data elements retrieved; 4) store the new data elements created by an analysis in

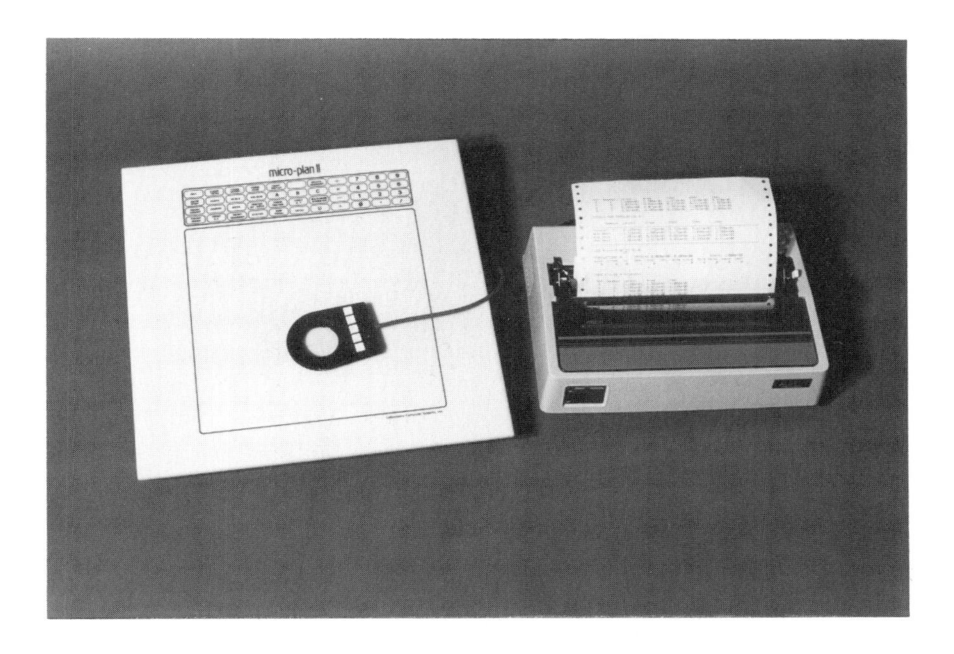

Fig. 48. Micro Plan II. This micro-processor based system can digitize individual points or a stream of points. It can also calculate length, area, and many other morphometric measurements from data traced on the digitizing tablet. Courtesy Laboratory Computer Systems, Inc.

the data file; 5) search, identify and route a variety of different data items and score these values with assigned weighted values (for example, search for optimal highway routing); 6) perform statistical analysis, such as multivariate regression, correlations, etc.; 7) overlay one file variable onto another, for example, census tract data over a land use map; 8) be capable of modeling and simulation, i.e., developing scenarios (generally in map form) to predict a future event [19].

7.3.2. *Remote Sensing Inputs*. The potential contribution of remote sensing to geographic information systems is enormous. Unfortunately, to date the actual contribution has been quite limited. Shelton and Estes [21] have blamed the limited progress in integrating the two systems on the fact that each was developed independently of the other. An annoying result is that data derived from remote sensing techniques are typically incompatible with most geographic information systems. Both Landsat digital data as well as data from conventional aerial photos must undergo complex, not to mention expensive, procedures before entry. Landsat digital data must be transformed geometrically; aerial photo data must be computer-coded and digitized (Fig. 48).

The process of transforming Landsat digital data involves the use of the appropriate computer software and was discussed in the previous chapter. As for the digitization or coding of aircraft data there are several techniques. The least expensive technique is manual digitization. This involves placing a transparent grid over a photograph or map overlay and recording the dominant class of data for each individual cell. Point and line data are recorded as vectors. The grid cell data and vector data are input to the computer either directly, at a teletype, or indirectly, by punching cards. Though a relatively simple and straightforward method, manual digitization is a very time-consuming one.

A more efficient method for encoding aircraft data is by semi-automated digitization. This method involves the use of an electromechanical digitizer to systematically record the x, y coordinates of landscape features. The photo or map is mounted on a digitizing table and the features are manually traced by the operator using a movable cursor. At regular intervals or at pivot points, x, y coordinates are recorded by manually positioning the intersection of the crosshairs on the cursor over the intended point and pressing the recording button. These vector data are stored in computer files or on magnetic tape. Automatic digitizers are also available. These devices sense and track linear features by means of servomechanisms. Operators must be present, primarily to resolve uncertainties posed by intersecting lines. Both systems are more expensive than manual digitization but they are also more accurate.

7.3.3. *Case Study*. The Minnesota Land Management Information Center (LMIC), a division of the state's Department of Energy, Planning and Development, is a depository and clearinghouse for natural resource information. As a part of its activities, LMIC operates and maintains the Minnesota Land Management Information system (MLMIS), an automated land resources information system serving both governmental agencies and the private sector on an equal basis [22].

The MLMIS System operates on a PRIME 850 minicomputer. A package of computer software called the Environmental Planning and Programming Language (EPPL) has been internally developed for the manipulation of the land resources data. The latter may be retrieved in tabular, statistical, cartographic and computer file formats. The system is also capable of handling Landsat digital data.

The software for MLMIS has been designed to handle any size grid cell. For the state as a whole the smallest geographic entity identified is the 40-acre parcel. This cell size was chosen because historically it was the basis for allocating land in the state. By the U.S. Public Land Survey begun in Minnesota in 1848 the state was divided into townships six miles square. Each township was further subdivided into 36 one-mile-square sections which were themselves subdivided into quarter sections. The latter were one-quarter mile on a side, or 40 acres. Though today many of the 40-acre parcels have been subdivided into a host of tiny parcels, features such as roads and certain property boundaries still reflect the 40-acre pattern. All 40-acre data in MLMIS are referenced to the UTM coordinate system.

Table 11. Data files maintained by MLMIS

Data files
Geographic data base
Township boundaries
Minor civil divisions
Public Land Survey description
School districts
Resource data
Land use/land cover
Water features
Roads and intersections
Soil types
Physiographic features
Forest types
Major and minor watersheds
Water sources for irrigation
Public land
Data items not in 40-acre grid format
Boundaries of census tract units
River and stream locations
Digital terrain data
Landsat data

The MLMIS data base is composed of two parts. The primary part is the 40-acre grid cell data. There are additional data, however, that have been entered in other formats including point, line, and polygon. Table 11 provides a list of some of the data files maintained by the MLMIS. The list is not intended to be a complete one.

Data sets maintained by MLMIS can be analyzed and displayed in a number of ways. In terms of the former, single variables may be manipulated or several sets may be merged for multi-variable analysis. Software has been developed that will also allow the user to evaluate the geographic relationship of one cell to surrounding cells.

Data output from MLMIS can be shaped to fit the particular needs of the user. Black-and-white maps may be generated from a lineprinter or an electrostatic plotter. Vector data may be displayed two- or three-dimensionally by a drum pen plotter in either black-and-white or color. As for displays on the high-resolution CRT, they may be converted to hard copy either through the use of an instamatic camera attachment or a film image recorder.

8. Land use/land cover: inventory and change

8.1 Introduction

Land use information is an important input to a host of planning decisions and for this reason has been gathered by various levels of government for many years. Existing land uses, because of their strong influence on how land will be used in the future, are a crucial element in the development of a land use or land development plan. They are also useful for purposes of tax assessment, for the development and implementation of zoning ordinances, and for various conservation activities. Perhaps most importantly land use data represent a base against which to measure change. Land use change data are necessary both for developing land use policies and for monitoring their effectiveness.

In recent years, the term 'land cover' has come to be commonly used in association with the term 'land use'. The two are not synonymous. Land use includes 'everything land is used for by residents of the country, from farms to golf courses, houses to fast food establishments, hospitals to graveyards' [23]. Land cover refers more to 'the vegetational and artificial constructions covering the land surface' [24]. Relying almost exclusively upon spectral data for recognition purposes, Landsat provides land cover data. Conventional aerial photointerpretation techniques, on the other hand, with their use of size, shape, relative location, and other photo elements, are better able to provide information on land use activities. But photos, too, have their limitations. Many land uses cannot be determined from the air because they depend upon land ownership. A typical example might be a state park used for hiking, picnicking and other recreational activities. On a photo, the park may appear as just another tract of woodland. Thus, land use inventories tend to be a mixture of land use and land cover data supplemented by various types of ground-derived information such as land ownership.

In spite of its limitations remote sensing has played an important role in the land use data acquisition process for many years. As early as the 1940's, black-and-white aerial photos were being used for this purpose. Since then, new films, like Ektachrome-Infrared, and new platforms, such as the U-2 and

Landsat, have added immeasurably to our means of acquiring these data. Over large areas in particular, remote sensing techniques have now become the single most effective method for land use/land cover data acquisition. For this reason, state planning organizations are quickly adapting high-altitude aircraft and Landsat techniques to their data-acquisition needs. As for local planners, with their requirements for data at individual parcel levels, they have continued to depend more heavily upon large-scale aerial photos and various ground-survey techniques.

8.2 Classification

Historically, land use data have not been gathered in any uniform manner. On the contrary, the various agencies acquiring data have tended to work quite independently of one another. The result has been a series of collection methods and classification systems having so little in common that data sets cannot be merged or aggregated. In an effort to improve this situation, the U.S. Geological Survey has developed a standardized system for classifying land use data obtained by means of remote sensing techniques such that detailed land use studies at local and regional levels can be aggregated upward to state and national levels. The system was developed to meet the following criteria [25]:

1. The minimum level of interpretation accuracy in the identification of land use and land cover categories from remote sensor data should be at least 85 percent.
2. The accuracy of interpretation for the several categories should be about equal.
3. Repeatable or repetitive results should be obtainable from one interpreter to another and from one time of sensing to another.
4. The classification system should be applicable over extensive areas.
5. The categorization should permit vegetation and other types of land cover to be used as surrogates for activity.
6. The classification system should be suitable for use with remote sensor data obtained two different times of the year.
7. Effective use of subcategories that can be obtained from ground surveys or from the use of larger scale or enhanced remote sensor data should be possible.
8. Aggregation of categories must be possible.
9. Comparison with future land use data must be possible.
10. Multiple uses of land should be recognized when possible.

The USGS classification system is comprised of four levels of categorization. Levels I and II of the system (Table 12) are expected to be utilized by planners at state and federal levels of government. Level I data will be acquired by

Table 12. Land use and land cover classification system for use with remote sensor data

Level I	Level II
1 Urban or Built-up Land	11 Residential
	12 Commercial and Services
	13 Industrial
	14 Transportation, Communications, and Utilities
	15 Industrial and Commercial Complexes
	16 Mixed Urban or Built-up Land
	17 Other Urban or Built-up Land
2 Agricultural Land	21 Cropland and Pasture
	22 Orchards, Groves, Vineyards, Nurseries, etc.
	23 Confined Feeding Operations
	24 Other Agricultural Land
3 Rangeland	31 Herbaceous Rangeland
	32 Shrub and Brush Rangeland
	33 Mixed Rangeland
4 Forest Land	41 Deciduous Forest Land
	42 Evergreen Forest Land
	43 Mixed Forest Land
5 Water	51 Streams and Canals
	52 Lakes
	53 Reservoirs
	54 Bays and Estuaries
6 Wetland	61 Forested Wetland
	62 Nonforested Wetland
7 Barren Land	71 Dry Salt Flats
	72 Beaches
	73 Sandy Areas other than Beaches
	74 Bare Exposed Rock
	75 Strip Mines, Quarries, and Gravel Pits
	76 Transitional Areas
	77 Mixed Barren Land
8 Tundra	81 Shrub and Brush Tundra
	82 Herbaceous Tundra
	83 Bare Ground Tundra
	84 Wet Tundra
	85 Mixed Tundra
9 Perennial Snow/Ice	91 Perennial Snowfields
	92 Glaciers

Landsat employing either image interpretation or digital processing procedures. Level II will be most efficiently and economically acquired by high-altitude aircraft such as the U-2, although it would also be possible to obtain them from medium- to low-altitude aircraft. Levels III and IV of the USGS system do not appear in Table 12. As a means of promoting flexibility, these categories are to be developed by various local planners to meet their own specific requirements. For the classification system to meet its objectives, however, the categories adopted by local officials must be capable of being aggregated into Level II categories. An illustration of this is provided by Table 13. It is anticipated that

Table 13. Residential land use categories at levels I, II, and III

Level I	Level II	Level III
1. Urban	11. Residential	111. Single-family units
		112. Multi-family units
		113. Group quarters
		114. Residential hotels
		115. Mobile home parks
		116. Transient lodgings
		117. Other

Level III data will be acquired from aerial photos of scales ranging from 1:80,000 to 1:20,000 while Level IV data will require scales of 1:20,000 or larger. Both Levels III and IV will call for considerable supplemental information.

8.2.2. *Landsat-Derived Data.* Land use/land cover data may be derived from Landsat products in two ways – from the imagery and from the CCT's. Data produced from the imagery may be compiled in map form for a manual information system, and/or digitized for input to a geographic information system.

Deriving land use/land cover information from Landsat imagery is done infrequently, though the information acquired in this manner may be quite accurate. For this type of mapping it is best to use computer-enhanced 1:250,000-scale Landsat images, preferably the false-color composites. Level I data can be acquired by employing conventional photointerpretation methods. Experience has shown this technique to work best in semi-arid regions where cultural features stand out in sharp contrast to the physical environment. In more humid regions, heavy vegetation tends to mask much cultural activity.

Early in the Landsat (ERTS) Program researchers at Dartmouth College employed conventional techniques to produce a land use/land cover map of southern New England from 1:250,000-scale false-color composites [26]. In spite of the heavy cover of forest and woodland, eleven categories of land use/land cover were identified. Of the eleven categories, the non-builtup ones (i.e. water, woodland, agricultural land and marshland) were most easily mapped. The builtup categories, on the other hand, were more complex and difficult to delineate. Since individual structures and streets were not visible, builtup categories were based upon density of development – commercial/industrial, residential high density and residential low density. Rural residences as well as those along the urban fringe were completely undetectable and as a result these areas tended to be categorized either as woodland or rural open land.

The use of Landsat digital techniques provides a more efficient, though complex, approach to land use/land cover mapping, particularly over large areas. Digital techniques typically involve several steps including preprocessing, classification, and output generation. Preprocessing refers to the various operations for removing radiometric and geometric distortions. Most of these operations are routinely conducted by the EDC although certain tasks, such as the reformatting of data for use with particular software systems, may have to be performed locally. Classification is the critical step since it is here that pattern recognition algorithms are applied to the preprocessed data in order to generate useful land use/land cover information. This information may be statistical and/or graphical in form.

Digital mapping techniques are based upon the assumption that different land uses and land covers have different spectral properties. By applying certain pattern recognition algorithms to Landsat data the various pixels may be statistically separated into clusters or classes defined by their spectral properties. This may require the use of two, three, or even four spectral bands. To determine these classes, training site data must be established either through the use of a supervised or an unsupervised clustering technique. These data are employed to set the spectral limits for the various classes. A number of pattern recognition algorithms, such as maximum likelihood or parallelepiped, are then available to assign pixels to the statistical classes into which they best fit according to the training site information.

The classification process often produces more classes of data than are needed. For instance, water may be classified as clear, turbid, and shallow although only one category may suffice. Thus, a number of classes may have to be merged. The merging process may be done through the use of aerial photos or ground survey data. It may be aided further by reference to the various statistics generated for each class during the classification process. These data include mean reflectance per spectral band, covariance matrix, inverse matrix, and divergence matrix.

8.2.3. *Aircraft-Derived Data*. The U.S. Geological Survey has a program underway to produce land use/land cover maps for the entire United States. The basic data sources for compiling the maps are high-resolution black-and-white mapping photography (1:80,000) and U-2 acquired color-infrared photography (1:60,000). Landsat data are used in a complementary manner.

Three products are being generated by the USGS program. One is a set of land use/land cover maps formatted either to the standard USGS 1:250,000-scale topographic maps or to the new USGS 1:100,000-scale series. Four associated maps accompany this set. They include political subdivisions, hydrologic units, census county subdivisions, and areas of federal land ownership. A second product is a map series consisting of 1:250,000-scale or 1:100,000-scale land use/land cover maps composited on a planimetric base. The maps are two-color with the land use data in black and the planimetric base in green. The third product is a set of land use/land cover statistics broken down by political units, hydrologic units, census county subdivision, and areas of federal land ownership. The data provided in these summaries have been generated by digitizing the land use/land cover polygons and inputting the data into the USGS's Geographic Information Retrieval and Analysis System

(GIRAS). The System converts the digital polygon data to grid cells of approximately 4 hectares (10 acres) and composites these data into summaries. GIRAS has the capability of plotting and replicating data in a variety of formats [27].

The classification system used for the land use/land cover maps consists of Level I and Level II categories. Additional categories at Levels III and IV may be added at the user's convenience. Most categories are mapped at a minimum parcel size of 16 hectares (40 acres). A few categories, including Water and Urban/Built-up, are mapped at a 4-hectare minimum. The program is expected to be completed by 1985. As products are completed they will be placed on Open file at USGS.

The USGS system has been developed with the specific intention of mapping the entire U.S. As a result it is not totally applicable to local planning agencies where land use data are typically mapped by means of manual methods and at Levels III and IV of the USGS classification system. For local planning agencies a more general set of procedures should be followed. They should include 1) determination of land use categories; 2) minimum parcel size to be mapped; 3) techniques for assuring uniformity and consistency in interpreting and mapping land use; 4) methods for assessing accuracy of data; 5) nature of the final product(s). Though some of these points have been discussed in previous chapters, they are sufficiently important to warrant repeating here.

The specific land use categories to be employed in any mapping project depend primarily upon the anticipated uses of the data. It is for this reason that the authors of the USGS classification intentionally left the determination of Level III and IV categories to local planners. There may be instances, for example, when the concern is entirely with various categories of open land; all other categories may simply be lumped into a single builtup category. On the other hand, the need might be for great detail in urban areas. Whatever the need, the categories selected should be ones that can consistently be identified from aerial photos and in particular from the photos available for the project. A full description of each category should be provided to those doing the interpretation of the photos to assure as much uniformity as possible in categorizing the uses [28].

The minimum size of parcels to be mapped is another determination that must be made in advance. The appropriate minimum is itself a function of several considerations such as the anticipated uses of the data and the minimum parcel size that can legibly be printed on the final map. Typically, smaller land use parcels can be identified on photos than are mapped so the scale of the final map is usually the deciding factor. On occasion two minimum parcel sizes may be used – one for urban land uses and a larger one for rural land uses. To assure uniformity, and particularly when several individuals are doing the interpretation, it is advisable to prepare a template of the approximate minimum parcel size [29]. Such templates are very useful at the beginning of a mapping project although they tend to be discarded after a few hours of interpretation.

The techniques for transferring data from photos to the final map are numerous but the simplest and most frequently used is the overlay technique. Land uses identified on the photos are delineated on overlays to the photos and numbered or colored accordingly. These data are then transferred either directly onto a map base or to a permanent overlay. The overlay material for the initial interpretation is usually frosted acetate or mylar which is fastened to the photos by means of a masking tape. Tick marks are typically placed at the corners and along the side for proper alignment. If data are to be transferred to a topo base, the borders of the topo sheet should be drawn on the appropriate photo or photos before interpretation begins. Otherwise more arbitrary areas may be delineated though they should be located in the center of photos to minimize the effects of relief displacement. It is best to use clearly defined linear features for boundaries wherever possible since they can easily and accurately be drawn on the adjoining photos. Cotton gloves are recommended when using this technique so neither the frosted surface of the acetate nor the photos will be soiled.

Once the preceding has been completed, the interpretation of land use and its delineation on the overlay may commence. What collateral information is available should be used in the interpretation process. It is best to begin the interpretation of land uses by delineating all linear features such as rivers and roads even though such information will already appear on the data base. Linear features not only help orient the interpreter but they allow for the easy blocking out of land parcels (Fig. 49). Obvious land uses should be categorized first with the more difficult decisions left to last. Those that cannot be identified from the photos will have to be field checked. But the interpreter may face other problems. For example, what should be done when several types of land use fall within the minimum size parcel? Typically, the predominant land use within the parcel is selected but if no single use predominates a mixed category may have to be used. Another problem involves land uses without clearly definable boundaries. Where zones of transitional change are found, a median line is often selected for the boundary. For the sake of consistency these issues must be settled in advance.

When the initial interpretation has been completed, the overlays must be edited and the accuracy of the land use interpretations assessed. All parcels intentionally left blank must be verified in the field. As for the interpretations it is generally too costly to verify them systematically so they are more commonly checked for accuracy on a sample basis. One simple way this can be done is to place a transparent grid over each overlay and fieldcheck the interpretations at the intersections of the grid lines. A percentage of accuracy is derived by

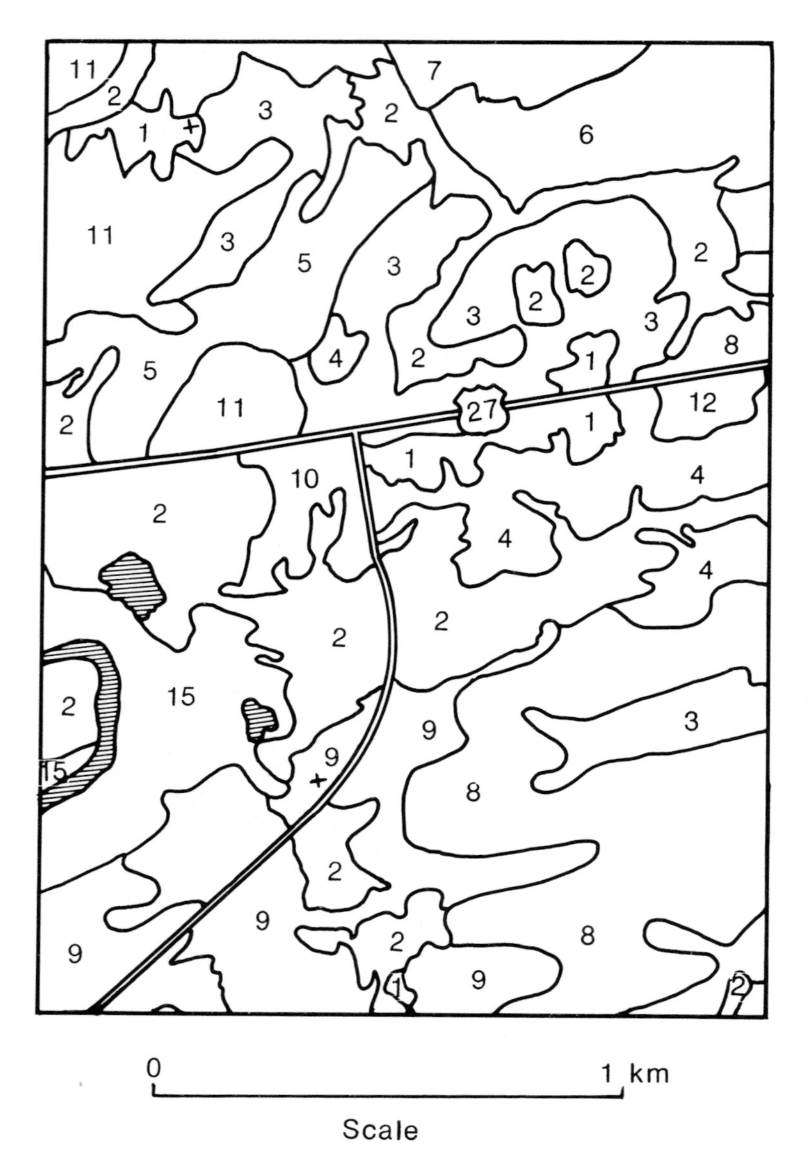

Fig. 49. Idealized land use map.

dividing the total number of sampled points by the number correctly interpreted. Anderson *et al.* has recommended that at a scale of 1:100,000 the accuracy should be at least 85 percent; at larger scales the percentage should be even higher [3].

Fig. 50. Aerial photo overlaid with a dot grid. A dot grid may be used as a means of estimating land use acreages or of assessing the accuracy of the interpretation.

8.3 Change Detection

Man's use of the land is not a static process but an extremely dynamic one. Changes in land use are taking place continuously in all categories though it appears that certain types of land use change such as the loss of prime farmland have received a disproportionate share of media coverage. An important reason then for inventorying land use is to provide a base against which to measure future changes in land use. There are four basic methods for detecting land use and land cover changes: original source material to new source material; original map to new source material; original map and source material to new source material; and original map to new map comparisons [31]. In the original source material-to-new source material change detection technique, changes are derived from a comparison of two sets of aerial photos. The photos representing different time periods are precisely registered to one another and changes in land use/land cover are identified and recorded. Ideally, the two photo sets should be comparable in terms of aerial coverage, scale, resolution, and time of day and year acquired. The less these conditions are met, the more cumbersome and labor-intensive this method can be.

In the original map-to-new source material procedure, land use and land cover changes are identified by comparing the original compilation with new (and rectified) aerial photographs. A transparency is made of the original compilation which is then overlain to the photos. Changes between the original compilation and the aerial photos are drafted directly onto the transparency. This is one of the least expensive and expedient of the four techniques.

The most comprehensive technique, and therefore the most recommended, is the original map and source material-to-new source material procedure. New aerial photos are compared not only to the original photos but the original map compilation which has been scaled to the new photos. Changes in land use/land cover detected on the photos can be precisely measured and mapped on the original compilation. An additional advantage to this procedure is that errors on the original compilation can be identified and corrected. On the other hand, if the two sets of photos are not comparable the process of change detection can be laborious.

The final method, original map-to-new map comparison, is probably the least recommended on the four. The procedure involves the compilation of an entirely new land use/land cover map, which is both expensive and unnecessary. Without reference to photos, compilation errors cannot be detected. Furthermore, changes representing nothing more than interpretation differences may also pass unnoticed. Whenever possible, both the original and the new photos should be consulted.

8.3.2. *Aircraft Techniques*. MacConnell and Niedzwiedz's study of Worcester County, Massachusetts, part of a larger study for the entire state, provides an excellent example of how aerial photos can be used to measure land use change [32]. In 1951 MacConnell had mapped land use and vegetation types from 1:20,000-scale black-and-white photography. Parcels of ten acres or more were identified and the information was transferred to 1:31,680-scale USGS topographic maps. Twenty years later, in 1971, the project was repeated. New black-and-white photography was flown at the same scale although this time parcels as small as 3 acres were mapped and transferred to 1:24,000-scale topographic maps. The use of USGS topographic maps for compiling the data was recommended because of the considerable detail already appearing on the maps. It was necessary only to update maps where new roads and major buildings had been constructed, and to transfer land use and vegetation data to them. The finished product was a USGS sheet with type polygons and type symbols added to it in India ink.

Acreages of the various land use types were computed by means of a dot grid. The classification system consisted of 104 categories. To simplify the computation process, these categories were aggregated into 28. For each quad sheet, an atlas of four hand-colored maps was produced; one displayed only the agricultural categories [5], a second the forest categories [7], a third the types of wetlands [4], and the fourth depicted the urban and miscellaneous categories [12]. Acreage determinations were made from these individual maps and compared with the 1951 data. The authors estimated that it required about two man-days to color each map and another two man-days to count the dots to determine the acreage by types. For the entire process – that is, for the photo-interpretation, cartography, and area measurement – it took approximately 150 skilled man-hours for each USGS topographic sheet.

Land use change mapping may also be done from high-altitude aerial photography. The Dartmouth College Project in Remote Sensing (DCPRS), under a contract from USGS, employed RB-57 acquired color-infrared photography to detect changes in land use for the metropolitan areas of Boston and New Haven [33]. Land use changes of greater than 10 acres were identified by location and type (26 categories) for the period 1970–1972.

To carry out the project the DCPRS was furnished, by the USGS, with a 9-plate land use map of the Boston area and a 4-plate land use map of the New Haven area; both sets were of a scale 1:62,500. The plates consisted of a black-and-white mosaic transparency with a mylar land use overlay. The land use overlay had been produced by the DCPRS under an earlier USGS contract. Both the mosaic and the land use overlay had been compiled from 1:100,000-scale color-infrared photography acquired in 1970. The DCPRS was also supplied with new (1972) 1:130,000-scale color-infrared photography.

Two methods were tested for detecting changes in land use: a side-by-side method and an over-and-under method utilizing a Bausch and Lomb Zoom Transfer Scope. The side-by-side method involved placing a 1970 photo adjacent to the comparable 1972 photo in order to facilitate viewing. A rectified grid was placed over both images to allow for orientation and the transfer of change data to the USGS mosaic. A piece of mylar was overlaid to the 1972 photo and the comparable USGS mosaic. Land use between the two periods was compared on a cell-by-cell basis. When a change in land use was detected, it was marked on the overlay to the 1972 photo and later transferred to the mosaic overlay.

With the side-by-side method, it was possible to complete one USGS plate every three days. The estimated land use change detection accuracy was approximately 80 percent. However, transferring these changes to the overlays of the USGS mosaics led to a decrease in accuracy to about 70 percent, due to the frequent absence of easily discernible registration points, such as highways, railroads, and rivers. Ponds and lakes were most often used for this purpose but there were too few of these to accurately locate every change in land use. A further complication was the difference in scale between the mosaics and the photos. The former were 1:62,500 while the latter were 1:130,000.

The second and more successful method for detecting land use change involved the use of the Zoom Transfer Scope. Specifically, a 1972 photo was placed on the photo easel and the comparable 1970 photo on a light table beneath the Zoom Transfer Scope viewer. A mylar sheet, on which changes were to be annotated, was placed over the 1970 photo; several water bodies were delimited on the mylar overlay for orientation purposes. The zoom magnification and anamorphic controls were adjusted in order to visually match the scales of the two photos.

Changes were detected by alternately increasing and decreasing the illumination behind the 1972 photo. Ground scarring, very visible on color-infrared photography, made detection relatively easy. When a photo was completed, the change overlay was placed on the photo easel and a blank mylar overlay to the mosaic took the place of the 1970 photo on the light table. The scale of the change overlay was made comparable to the mosaic by aligning landmarks such as lakes and ponds. Changes were then transferred to the overlay to the mosaic.

The use of the Zoom Transfer Scope enabled one full plate to be completed each day. Just as importantly, the estimated change detection accuracy increased to 85–90 percent and the locational accuracy to approximately 85 percent. Locational inaccuracies could not be entirely eliminated because of the differences in the scales of the two sets of photography. The 1970 photos were 1:100,000-scale while the 1972 photos were 1:130,000-scale.

As a final step in the change detection process, the coded land use maps and change overlay were manually digitized using a grid-cell technique. Each cell represented an area of 1/25 square kilometer. Grid-cell data were acquired for each of the 9 plates covering the Boston area and the 4 plates covering the New Haven area, and were typed into the computer. A file of town boundaries was also created with the same format and scale of the data bases. Tabular and map data on change were then generated for whole areas and for each individual town within the areas.

8.3.3. *Digital Change Detection*. The detection of land use/land cover change is a complex procedure which as yet is incapable of producing consistently accurate results. Overall accuracies average about 70–75 percent, but for some individual categories the accuracy rate may be as low as 50 percent or less. The reasons for these low figures are numerous: the difficulty in precisely registering or overlaying multiple sets of Landsat data, the complications posed by atmospheric haze, the lack of sufficient spectral separability between various categories of land use, and the complexity of land uses within certain areas, most notably, urban ones. Yet in spite of these problems, efforts to improve digital change detection techniques continue. As accurate as aerial photointerpretation data may be, they are extremely time-consuming to generate and difficult to replicate [34].

Digital change detection is based upon the assumption that a change in the use of a particular parcel of land will accordingly lead to a change in the spectral response of that parcel. Thus, for example, a land parcel used as a woodlot at one period would display a very different spectral response at a second date if in the meantime the woodlot had been replaced by a subdivision. Ideally, a map displaying pixels where a change in spectral values had taken place between two periods could be used to identify 1) the location, 2) the type (by the nature of the spectral change), and 3) the amount of land use change. More realistically, the change map probably shows:

a) large contiguous areas of spectrally discernible land use/land cover change;
b) large contiguous areas of land use/land cover change not spectrally discernible and, therefore, not mapped;
c) small noisy parcels incorrectly mapped as change due to effects of misregistration, edge pixels, and so forth;
d) small parcels of actual land use/land cover change not mapped due to misregistration, spectral separability, sensor spatial resolution, or other factors [35].

It should be emphasized that the MSS system was designed less for land use/land cover mapping purposes than for geologic applications. There are only four spectral bands for which data are acquired and these four bands tend to be redundant. Bands 4 (0.5 to 0.6 μm) and 5 (0.6 to 0.7 μm) are the only two visible bands. Band 5 is superior for identifying cultural features; band 4 adds little information. Bands 6 (0.7 to 0.8 μm) and 7 (0.8 to 1.1 μm) are reflective-infrared bands. They both provide information on vegetation and water resources, though band 7 seems to be superior for this purpose. Here again the Thematic Mapper would appear to offer considerable promise. Not only does it have a smaller pixel size than the MSS, it acquires data in several spectral bands, although one of these was selected for geologic purposes. It is fortunate that change detection requires less spatial and spectral resolution than does the initial identification of an object [36].

A number of algorithms are used for change detection analysis, including image differing, image ratioing, classification comparison, and change vector analysis. The decision as to which algorithm is most appropriate to use rests on several factors. For one, a familiarity with the physical and cultural characteristics of the area under study is essential. It is also important to know how precisely the two data sets have been registered. Finally, one should be aware of the various change detection algorithm alternatives, their degree of flexibility, and their availability [34]. Only two of the more commonly employed algorithms will be described here – image differencing and image ratioing.

Image differencing is a method for calculating the difference in reflectance values between two sets of Landsat data on a pixel by pixel basis. To accomplish this the two data sets must first be geometrically corrected, scaled, and registered. Then, the digital values for each band in one data set are subtracted, pixel by pixel, from their counterparts in the other set. Precise registration of the two sets becomes crucial here. The subtraction process results in positive and negative values in areas of change. (Negative values are usually converted to positive by the addition of a constant.) Where no change has occurred, values should approximate zero.

The image differencing procedure produces values for each band that are approximately Gaussian in distribution. Pixels of no radiance change are distributed around the mean while pixels of considerable change are distributed at either end of the curve [37]. The critical element in the method is determining where to place the threshold boundaries between the change and no-change pixels displayed in the histogram. Typically a standard deviation from the mean is selected and empirically tested to see if changes were accurately detected [34].

When displaying the differences graphically, small differences between the two data sets are assigned medium gray tones; larger differences are expressed in progressively lighter or darker tones, depending upon which data is subtracted from the other. Color composite images can also be produced utilizing difference images for any three of the four spectral bands [19].

Image ratioing is a method of change detection that attempts to compensate for the differences in sun angle, sunlight intensity, and shadows that may exist between data sets of different dates. Presumably, fewer inaccuracies will occur in change detection data if these factors can be controlled.

The image ratioing procedure is similar to image differencing except that radiance values of two data sets are ratioed rather than subtracted from one another. Like image differencing the two data sets must first be geometrically corrected and registered. Then the radiance values for one data set are divided by the radiance values for the second data set, pixel by pixel. The resulting quotients are new numbers that theoretically can range from 0 to infinity, though in practice they tend to range from 0.3 to 3. Because the quotient values fall within such a narrow range relative to brightness values, the ratio numbers are usually expanded to a range of 0–255. Again, as in the case of image differencing, the threshold limits for determining change are based upon empirical judgments.

Images resulting from the ratioing technique display tonal contrasts only where change has occurred. Theoretically, differences caused by shadows and changing sun angle have been eliminated. Maps produced by this method are actually maps of autocorrelation or redundancy between data sets. Unfortunately, this technique has been criticized as being statistically invalid [38].

8.4 Conclusion

For the foreseeable future, aerial photos will remain the primary tool for operational land use/land cover inventory and change detection programs. Compared to Landsat digital techniques, aerial photointerpretation techniques are more flexible and generally less expensive to employ over small areas. They also provide more accurate data. Landsat digital processing techniques are not yet capable of consistently detecting land use/land cover change in a given area with a high degree of accuracy (greater than 85 percent) for all standard classification categories. For some categories, accuracy may be as low as 50 percent [39]. Then, too, since land use planning remains primarily a local responsibility, large-scale data are frequently required. This represents another reason for the primacy of aircraft data. While the Thematic Mapper should provide more accurate and larger scale land use/land cover data than Landsat's Multispectral Scanner, it will complement rather than supplant aircraft as the major data source.

9. Resource preservation

9.1 Introduction

The economies of many regions depend heavily upon their land resources. The most obvious of these resources are farmland, range and pasture land, and forest land. Proper management of land resources is mandatory if local economies are to remain healthy and living levels are to continue rising. Unfortunately the management of land resources is an extremely complex task. Successful policies must take into consideration two important concepts – highest and best use, and multiple use. Highest and best use refers to the tendency of landowners to allocate land to the use promising the highest return. In general industrial or commercial uses earn a higher return than other uses [40]. Multiple use refers by definition to the use of land for several (and on occasion conflicting) purposes. A tract of woodland, for example, may be used simultaneously for recreation, water protection, and wildlife habitat, as well as for producing timber.

Both the highest and best use and the multiple use concepts have become more important as metropolitan areas have continued to expand into rural areas. The latest census has revealed not only that the U.S. population is growing at a rate of approximately 2 million persons per year, but that between 1970 and 1980 rural areas grew as fast as urban areas for the first time since 1790 [41]. The growth of rural areas has been hastened in part by the completion of the interstate highway system. Areas previously accessible only with great difficulty have now begun to attract residential development, recreational facilities, and even industry. In turn, this growth has led to the loss or threatened loss of considerable open land. As the need for the more effective management of many of these land resources has increased, so too has the potential role of remote sensing.

9.3.2. *Farmland.* One of the nation's most valuable resources is its farmland, and the preservation of this farmland has become an important, though controversial, task. There are approximately 380 million acres of prime farmland in the United States, but according to an SCS study, there was a net loss of 30 million acres of farmland between 1967 and 1975 [42]. Much of the lost farmland was among the country's most productive. Unfortunately, many of the same qualities that make land prime for agriculture – level land, well-drained soils, dependable supply of water – also make land prime for urban development. Of the 30 million acres of farmland lost, an estimated 16 million acres were converted to urban uses (Fig. 51). Stripmining is one of the other activities responsible for the loss of so much farmland. In Illinois alone, 170,000 acres of farmland have been lost in this way [23].

Critics of the National Agricultural Lands Study contend that the estimates of farmland loss due to urban development are too high. At least one claims that much of the difference between the 1967 and 1975 figures can be attributed to a change in the methods employed to derive these figures [43]. While the dispute has yet to be settled, the study has made an impact. For one, Congress has enacted the Land and Water Conservation Act (1976) which requires among other things that the Secretary of Agriculture assess the state of the nation's land and water resources every five years. Many states have also adopted their own measures to preserve farmland. A total of 45 states have now enacted such legislation.

A number of techniques for preserving farmland are available to state and local authorities including agricultural zoning, transfer of development rights, differential tax assessment, and tax deferral. Whichever of these techniques is applied, an inventory of agricultural resources is a necessary first step. One source of such data is the SCS which has established a classification system composed of eight land capability classes, the first four of which are suitable for cultivation. The eight classes may be summarized as follows:

Class I – have few limitations that restrict their use. Nearly level; low erosion hazard; deep generally well-drained and easily worked soil.

Class II – have some limitations that reduce the choice of plants, or require special conservation practices, or both.

Class IV – have very severe limitations that restrict the choice of plants, or require very careful management, or both.

Class V – have little or no erosion hazard but have other limitations impractical to remove that limit their use largely to pasture, range, woodland, or wildlife food and cover.

Class VI – have severe limitations that made them generally unsuited to cultivation and limit their use largely to pasture or range, woodland, or wildlife food and cover.

Class VII – have very severe limitations that make them unsuited to cultivation and restrict their use largely to grazing, woodland, or wildlife.

Class VIII – soils and landforms have limitations that preclude their use to recreation and wildlife, or water supply, or to esthetic purposes [44].

The Vermont Department of Agriculture has developed a simple and inexpensive method of combining SCS soils data with existing farmlands data to

Fig. 51. Conversion of farmland to urban uses. Upper photo shows portion of Lebanon, New Hampshire, in 1937. Original scale 1:12,500. Courtesy U.S. Army. Lower photo shows same area today. Farmland has given way to a major interstate highway, shopping plaza, and several fast-food restaurants. Photo courtesy of Dan Baum.

produce a basic information system for farmland preservation activities [45]. The data base for the system is a series of 1:5000-scale orthophoto maps that have been produced for each town in Vermont by the State's Division of Property Valuation and Review. To this base is first added information on prime agricultural soils, available from District Soil Conservation Offices in the form of soil survey sheets. The information on these sheets is transferred to a transparent overlay either by 'eyeballing', of more obvious tracts, or by utilizing a zoom-transfer scope. On the overlay prime soils are labelled either 'H' for high potential or 'G' for good potential. High potential soils are the equivalent of Class I soils in the SCS classification system, good potential soils the equivalent of Classes II and III.

To the soils overlay is added a second on existing farmlands. All agricultural land uses are plotted whether they are found on areas of prime soils or not. This information is acquired from recent aerial photos with some checking done in the field. Also annotated on the overlay are the existing uses for prime soil areas not in farming. Such uses are classified according to Level II of the USGS system. A third overlay is being produced showing how prime soils areas were being utilized in 1939. The data are being derived from 1939 aerial photos and will show how the use of prime farmland has changed between 1939 and the present.

The information system being developed by Vermont's Department of Agriculture represents a visual inventory of the agricultural soils base of each town in the state. To this inventory towns may add information on land ownership, tax assessments, zoning and other restrictions. Such a system can be very useful in preserving farmland and avoiding potential land use conflicts. Many farmers have halted operations because of conflicts with new neighbors over farm odors, dust and noise.

To this point, the term 'preservation' has been used in the sense of keeping farmland from being converted to other uses, but it may also refer to the maintenance of soil quality. Maintenance of farmland quality means guarding against the salinization or flooding of soils from poorly designed irrigation systems, the prevention of erosion which may result from improper cultivation techniques or crop selection, and minimizing the impact and spread of insect and wildlife infestations. Unlike the farmland conversion process, which in most cases may only require new data every two or three years, the process of maintaining farmland quality requires continuous data.

The detection of soil problems, particularly over large areas, may be facilitated through the use of aerial photos but it is only one of many techniques that may have to be applied. Others may include ground observations, soil moisture measurements, water quality analysis, and plant tissue analysis to name just a few [46]. Nevertheless, aerial photointerpretation may speed up the detection process and later may provide an effective method for evaluating the treatment process and the overall farm management program.

The literature on the uses of remote sensing for farmland management is voluminous and the best single source of this information is the American Society of Photogrammetry's new *Manual of Remote Sensing* (1983). It is generally concluded that an effective farm management program must be based upon a detailed soils survey. For many areas of the United States, these data are available from either the Soil Conservation Service or the Department of Interior's Bureau of Land Reclamation. Even in the absence of such information, improper farming practices may be detected by means of remote sensing. In farming areas, survey companies are becoming increasingly available to provide this service.

Plants are good indicators of general soils conditions because their roots tap such large subsurface areas [47]. Distortion of plant shape, discoloration of foliage, and defoliation are all signs of plant stress and that stress may be the result of nutritional disorders, drainage problems, insect infestation, or disease. Color films are best for the detection of stress with color-infrared used most extensively. Color-infrared is preferred because of its sensitivity to changes in the reflectivity of near-infrared wavelengths. Any change in color either from red to pink or red to dark red and black may be indicative of stress. Experiments have shown that stress may also be detected on color-infrared film one to three days before visual symptoms become apparent [48].

Causes of plant stress and therefore farmland management problems may be grouped into three categories: nutritional disorders, drainage problems, and disease or insect infestation.

Nutritional disorders typically result from the uneven application of chemical fertilizers. Whether custom-applied or applied by the farmer, missed strips or strips where the fertilizer was insufficiently spread become increasingly visible from the air as crops mature. On occasion, less regular patterns may indicate areas where a different mix of chemical elements is required. Nitrogen-deficient plants, for example, display increased reflectivity in both the visible and infrared wavelengths – in the visible because nitrogen-deficient leaves have less chlorophyll, in the infrared because they have thicker leaves [49].

Plant stress caused by drainage problems implies that plants are receiving too much or too little water. The former often causes plants to produce more vegetation but less fruit; the latter, too little water, may be fatal. Differences in water supply may be related to the design or the operation of an irrigation system. There may simply be gaps in the system which can be corrected by moving sprinklers or pipelines closer together. Worn sprinkler nozzles have been found putting out 200 more gallons of water per revolution than planned [50]. On the other hand, differences in water supply may result from soils variations. For example, coarse-textured soils have less of a water-holding capacity than fine-textured soils. As a result, a water application rate calculated on the basis of coarse-textured soils may be too great for areas of fine-textured soils. Any accumulation of water in these latter areas may drastically reduce yields and possibly foster the spread of certain diseases.

A related problem is soil salinization. Where the application of water is excessive, an increase in soil salinity may result. If the level becomes high enough it will be fatal to vegetation. Generally, as salinity increases leaf color and leaf thickness are affected. Cotton plants subjected to increased soil salinization typically appear darker on color-infrared photography than normal appearing plants [49]. The only cure for highly saline soils is to flush them with clean water or install tile drains if the water table is too high for flushing [46].

The detection of disease and insect infestation is a task requiring constant attention and over large cropland areas there are few alternatives to aerial surveys for doing so. Generally, surveys must be carried out quickly, either because in the case of insects, they may have a short life-cycle, or, as occurs with certain diseases, the plants may be dead within a week or two of the initial infection. Depending upon the crop type being monitored, color or color-infrared would be the preferred film. Once an area of plant stress has been detected, field analysis should quickly reveal the causal agent. For both financial and environmental reasons, application of the appropriate insecticide or fungicide may then be concentrated only on the infected area. Further monitoring will be required to ascertain that the spread of the disease or insect infestation has been halted. If done early enough in the season, damaged areas may be replanted. If the disease or insect infestation persists over two or more growing seasons it may become necessary to rotate crops before complete eradication takes place.

9.3.3. *Rangeland.* Nearly 50 percent of the earth's surface is covered by rangeland (which is here meant to include prairie, savanna, steppe, and llanos). In the United States rangeland, grassland, and pasture account for nearly 680 million acres. Though vast, such areas are characterized by marginal amounts of rainfall and extremely fragile ecosystems. Accordingly, rangelands must be managed with care. Complicating the management process is the vast number of functions rangelands must perform. Collectively they support millions of cattle, sheep, and goats as well as a considerable number of wildlife species. Rangelands are also utilized for cropland, watershed management, recreation, and increasingly even residential development. During cycles of less than average rainfall, many of these uses must be restricted for rangelands may suffer over-grazing, excessive erosion, and desertification. A classic example of this occurred in the African Sahel during the early 1970's when several years of drought with subsequent overgrazing caused the Sahara Desert to expand its borders south-ward into the savanna grasslands.

Rangeland management is based primarily upon ecological principles whose main components are soils, vegetation and animals. These components must be continually monitored in order to determine whether management practices are contributing to the long-term health of the range or to its deterioration [51]. Changes in plant-species composition or in the characteristics of soils are indicators of changing range conditions which normally require a reassessment of grazing practices. Traditionally, rangeland information has been gathered through field surveys and the detailed analysis of sample plots; increasingly remote sensing is playing a more important role. Once a solid information base has been established, remote sensing techniques can provide an effective substitute to ground surveys for the detection of changes in rangeland ecology.

Rangelands are vast areas and for much of the world Landsat may presently be the only way to gather soils and vegetation data. At Landsat scales (1:250,000) soils and vegetation are not delineated separately but in association with one another. The spectral signatures used in the delineation of soils/vegetation associations are themselves composites reflecting such factors as the type and density of vegetation, slope and elevation, rock type and/or soils [52]. Typically, supervised digital classification techniques enable land cover to be mapped at Level II of the Anderson classification system. But recently, investigators have found that when landform and land-cover ground truth data are used in conjunction with the digital imagery land-cover units in rangeland areas may be classified consistently to Level III and at times even to Level IV [53].

Zonneveld has argued that the best tool for large-scale rangeland surveys is superwide-angle black-and-white photography [54]. Such photos are scales of 1:50,000–1:70,000 provide an accurate means of depicting ecological areas because they combine high resolution with a stereoscopic capability. Neither this photography nor Landsat is capable of providing information on individual plant species or soils characteristics.

Spiers has described how 1:18,000 and 1:10,000 black-and-white photos were employed to map pasturelands in southwest Spain, a grazing area where government policies were encouraging the degradation of the region's vegetation cover [55]. The objective of the mapping project was to depict the vegetation cover in relation to a number of environmental variables including soils, topography, and moisture. Specifically, the photos were used to map vegetation and to select sites for field surveys. Although the complexity of cover types was great due to the intermingling of natural vegetation with culturally-induced species, the photos provided sufficient detail for delineating vegetation types in terms of species associations. The photos did not prove adequate for either species identification or the determination of grazing practices on individual pastures. The author suggested that 1:5,000-scale color transparancies would be necessary for acquiring the latter information. Colwell concurs that 1:5,000 is the best scale for rangeland surveys but recommends either the use of color-infrared film or black-and-white film with a light red filter [56]. Table 14 provides a summary of scales and suggested rangeland uses.

Table 14. Scales of remotely-sensed data used by rangeland managers with suggested rangeland uses*

Very large-scale	1:100–1:500	Species identification including grasses and seedlings, erosion estimates, rodent activities, assessing surface soil factors including litter.
Large-scale	1:600–1:5000	Species measurements, erosion estimates, productivity estimates, condition and trend assessment.
Medium-scale	1:6000–1:32,000	Detailed vegetation mapping, rodent activities, erosion features. Condition and trend assessment. Vegetation mapping of plant community, larger erosion features, some planning within allotments.
Small-scale	1:40,000–1:125,000	Planning for a range management, vegetation and soil unit mapping on a pasture and/or allotment basis, multiple use planning.
Very small-scale	1:200,000–1:1,000,000	The synoptic view for planning rangeland use within the multiple use framework, mapping vegetation by zones covering large areas such as mountain ranges.

*Reproduced with permission from *Manual of Remote Sensing,* Second Edition, copyright 1983, by the American Society of Photogrammetry.

Water is another critical variable in rangeland management. Bodies of water provide breeding grounds for wildfowl, watering sites for animals both domesticated and wild, recreational areas for hunters, fishermen, and boaters, and sources of irrigation water for cropland. Because so many of these uses conflict with one another, water management is complex and requires continuous information on use activities.

Information on the location and extent of water bodies can be acquired from even small-scale imagery including Landsat. Water bodies can be depicted best on the infrared bands (MSS 6 and 7). Use activities require much larger scales. Deuel and Lillesand have suggested that 70 mm photography may be an inexpensive method for obtaining such information [57]. Employing 1:15,000-scale color transparencies, data were acquired for several boat types, boat moorings, and swimming rafts. They concluded that the method was not only less expensive but less complex than comparable field survey methods. The authors added that by increasing the scale to 1:10,000 and using stereo, far more boat types and activities could be identified. It can be inferred from this statement that the same large scale photos used for species identification could also provide information on numbers of hunting blinds, fishing craft, and even numbers of cattle or wildlife species found in the vicinity of water bodies.

Man's activities are not the only threats to rangeland ecosystems; nature provides its own share. Ants, rodents and fire are just a few of the agents responsible for destroying millions of hectares of rangeland vegetation each year. Rangeland managers must constantly monitor areas under their jurisdiction to detect vegetation damage, determine the causal agent, and introduce the appropriate abatement procedures. Further monitoring is necessary for evaluating the effectiveness of the abatement procedures as well as any subsequent reseeding efforts.

Early detection of damage-causing agents is critical both in terms of confining the agent for control purposes and for keeping damage to a minimum. One such agent that has so far defied all attempts to control it is the imported fire ant (*Solenopsis invicta*). Now infesting an area from Texas to North Carolina, the fire ant destroys an estimated $100 million worth of agricultural crops each year. In addition, it wrecks harvesting machinery and disfigures farmland, lawns, golf courses and parks with its mounds. Research conducted on the coastal plain of Texas has demonstrated that fire ant mounds can be visually detected on large-scale (1 : 2000) color, color-infrared, and black-and-white infrared photos [58]. The two infrared films proved superior to conventional color for detecting the mounds during periods of the year when there was little vegetation. During summer when vegetation begins to cover the mounds, color-infrared proves superior to black-and-white infrared. Of the several dates of imagery analyzed in the project, December was found to be the optimum month for detecting fire ant mounds in Texas. Similar techniques have been used to determine the level of activity of other damage-causing agents. Watson produced accurate estimates of gopher numbers from the presence of mounds revealed on 1 : 6000 color and color-infrared photos [59].

The use of Landsat has received particular attention because of its applicability to the vast and largely unmapped rangelands of Africa and the Middle East. Already Landsat is being used as an input to the FAO's locust habitat mapping program [60]. The locust lives in scattered locations from West Africa to India. Periodically, there occurs a vast upsurge in the numbers of locust to plague proportions. The resulting damage to crops and rangeland vegetation can be enormous. The upsurges in the locust population tend to occur after prolonged periods of heavy rainfall when vegetation in the breeding areas becomes well developed. The purpose of the mapping program is to use desert vegetation biomass as an indicator of potential locust breeding activity. Over such a large and inaccessible breeding area Landsat is the only source of biomass data. Ideally, when Landsat data display increases in vegetation density, those breeding areas can be singled out for locust control procedures. The same technique

Fig. 52. Harvester ant mounds. Photo is a black-and-white copy of a color-infrared image. Courtesy William G. Hart.

is also being used in Australia where McCullock and Hunter have demonstrated that visual analysis of 1 : 1 million Landsat color composites can be as effective in assessing the condition of locust habitats as presently employed ground survey methods [61].

Fire has long been an important variable in the management of rangelands. Whether caused by lightning or set by man, fire has altered the vegetation cover by increasing the grass cover at the expense of trees. In the savannas of Africa grass fires occur annually during the dry season (October–February). Landsat has enabled researchers for the first time to view the use of fire over large areas [62]. Apparently, herdsmen set many fires at the beginning of the dry season. Gradually these small fires expand and coalesce. By early February vast areas have been burned and some fire fronts are 6–10 kilometers in length. The charred areas show up clearly on Landsat imagery, although by the peak of the dry season smoke and haze may obscure much of the landscape.

A final requirement of rangeland managers may be a livestock census. For livestock numbers high-resolution black-and-white panoramic photography at scales of 1 : 12,000–1 : 15,000 will be sufficient. For determining the breed and class of livestock scales must be at least 1 : 8,000 [63]. Since shadows play an important role in the identification process, photos should be taken early in the morning or late in the afternoon. These two periods are also the cool part of the day when livestock tend to be feeding. As for the time of year, late spring is best since the weather is cool and there is not as much foliage to mask the animals.

9.3.4. *Forest and Woodland Management*. Forests and woodlands are one of man's most important resources. Although historically forests have been exploited as a source of fuel and construction materials, these are not the dominant concerns of forest management today. Just as important are the roles forests play in providing outdoor recreation, wildlife habitat, environmental amenities and water [64]. In addition, forest industries are also a source of jobs. In states such as New Hampshire, forest industries employ nearly 15 percent of the manufacturing work force while thousands more are employed in related service industries.

The most essential function of forest management is forest protection. Unfortunately, over vast areas of the earth man's activities have either already destroyed the forest cover (China and the Middle East) or, as in areas of Africa and South America, are in the process of destroying the forest cover at this very moment. Forest management requires the careful planning of harvesting operations and the protection of forests from fire, disease, insects, and even certain recreational activities. Management policies must attempt to satisfy the legitimate demands of a number of forest users.

Aerial photointerpretation techniques have been employed by forest managers for over half a century. They have proved useful for most forest management activities. Accordingly, the literature on the application of aerial photointerpretation and remote sensing techniques to forest management is extensive and only a most cursory review will be presented here.

Proper forest management requires an inventory of forest resources. Typically this means the identification and mapping of tree species or associations of species. The extent to which species identification can be accomplished from aerial photos is primarily a function of photo scale. At very large scales (1 : 3,000–1 : 5,000) tree species can be readily identified by crown shape and branching patterns. As photo scales decrease (1 : 15,000) it becomes difficult to distinguish individual trees unless they are in the open. At this point, tree shadows may become the important means of identification. At scales smaller

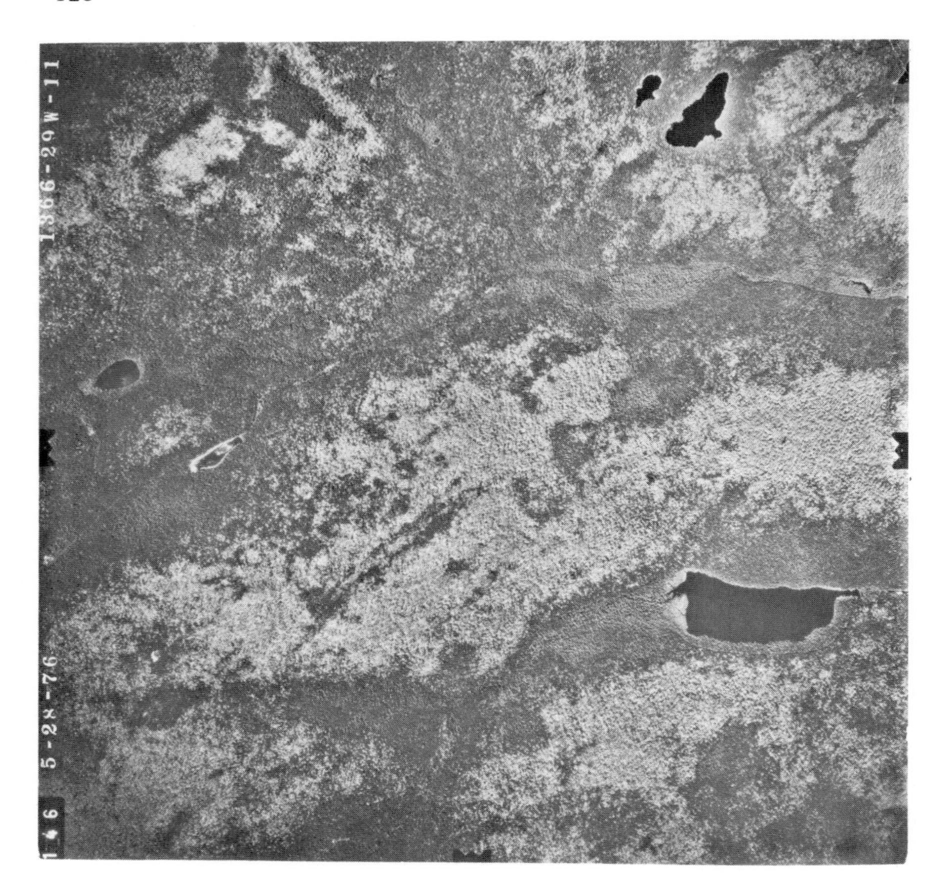

Fig. 53. Black-and-white infrared photo of forestland in northern Maine; scale 1:16,000. Light colored areas are deciduous trees (maple-beech-birch) while darker areas are spruce-fir. Intermediate tones along streams are northern white cedar-black spruce. Courtesy James Sewall Company.

than 1 : 15,000 tone and color become the primary means of identification [65]. Species associations may be mapped from photos of scales as small as 1 : 60,000 though this tends to be an easier task in northern forests than in temperate forests. Northern forests are characterized by few species while temperate forests have a great mix of species which are seldom found in pure stands (Figs. 53 and 54).

Species identification can be carried out on black-and-white photography although the degree of accuracy may depend upon a number of factors – the scale, quality, and season of photography, the film-filter combination, and the experience of the interpreter [51]. Black-and-white prints are typically preferred for use in the field since they hold up better than color prints. Nevertheless, the use of color photography in forest mapping is increasing. Both color and

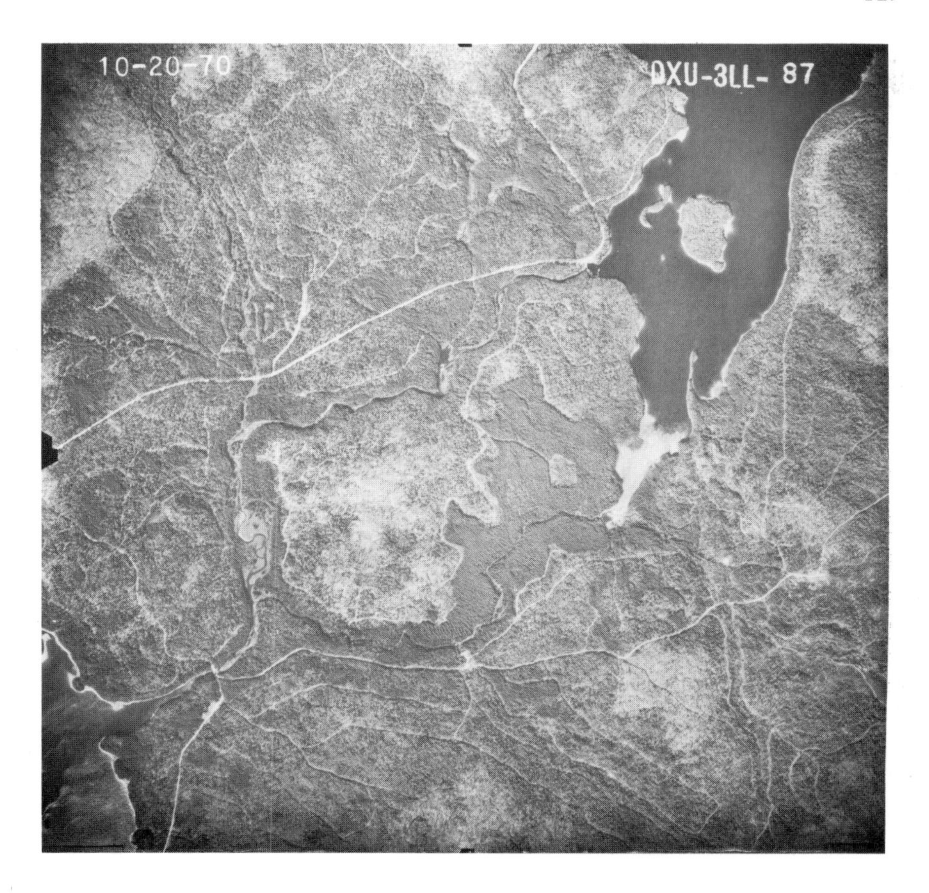

Fig. 54. Black-and-white photo of woodlands in northern New Hampshire; scale 1:20,000. The dark areas in the lowlands are spruce-fir; the lighter areas (uplands) are maple-beech-birch. Courtesy USDA.

color-infrared photography provide more information than black-and-white. Color photography at scales of 1:15,840, 1:20,000 and 1:24,000 is now routinely acquired over national forests by the U.S. Forest Service. At smaller scales color-infrared is often the preferred film. Not only does color-infrared provide an abundance of hues and colors for species identification, but the images display greater sharpness due to color-infrared's natural haze-penetration capability.

The great size of many forest tracts has led a number of investigators to experiment with the Landsat digital tapes as a possible source of forestry data. In general, experiments have found Landsat tapes capable of providing reasonably accurate data at Levels I and II of the USGS classification. As one example, Dodge and Bryant used a supervised classification of MSS data to classify several thousand acres of northern New England forest into one of five categories (pure hardwood, pure softwood, 50–50 hardwood/softwood, 75–25 hardwood/softwood, and 25–75 hardwood/softwood). They found that the acreages derived from Landsat data for each of these categories compared favorably with the estimates derived by the Forest Service [66]. At Level III results obtained from Landsat digital processing techniques have varied greatly among researchers. Middleton *et al.* found Landsat-derived acreage estimates for three forest types (pine, oak/pine, and oak/hickory) to differ from ground survey data by no more than 6 percent for any category [67]. Smit, on the other hand, found Landsat incapable of accurate differentiation between spruce, beech, and mixed species [68].

For most forest management activities, Landsat-derived data appear to lack both sufficient detail and sufficient accuracy. But perhaps just as importantly Landsat might not provide any savings in money either. Comparing 1:10,000-scale orthophoto maps made from 1:60,000 and 1:30,000 aerial photos with Landsat digital printouts Stellingwerf concluded that the Landsat printouts could not match the orthophoto maps for accuracy or detail though the costs for producing the two were about equal [69].

For harvesting purposes it is necessary to have information on species types but also on timber volume. Traditionally estimates of timber volumes have been derived by ground survey methods or cruising. Such methods provide accurate measurements of tree diameter, form class, and stem defects within preselected sample areas. Aerial photos are often used to improve the effectiveness of ground cruising by stratifying the field samples to include various tree species and stand-size classes [70].

Timber volumes may also be estimated from aerial photos. The accuracy of the estimates will depend upon such factors as the scale and quality of the photos, the experience of the interpreter, and the quality of the measurement devices [51]. Volume estimates (cubic foot or board foot) for a given area are derived on the basis of several measurements made directly from the photo. The usual components of volume estimates are tree height, crown diameter, crown closure, crown area, and number of trees. Average tree height is the component best related to volume and is derived either from shadow lengths or parallax measurements. Crown diameter is related to stem diameter and is used to determine size classes, i.e., pole timber, small saw timber and large saw timber. Crown closure or the percentage of an area covered by the tree canopy is another measure for the number of trees per unit measure. Volume tables have been developed for many important tree species and associations so volume estimates can be generated fairly quickly once measurements have been made of the various volume components. Table 15 represents an idealized volume table.

A less expensive method of estimating the area occupied by various species types is sampling. This is accomplished by placing a transparent grid over each individual photo and recording the species type at the grid line intersections.

Table 15. Idealized stand volume table in cubic feet per acre*

	Crown closure (%)				
Stand height (ft)	20	40	60	80	100
40	600	700	800	900	1000
50	700	800	900	1000	1100
60	800	900	1000	1100	1200
70	900	1000	1100	1200	1300
80	1000	1100	1200	1300	1400
90	1100	1200	1300	1400	1500

*Adapted from T. Eugene Avery, *Foresters Guide to Aerial Photo Interpretation*, 1966.

The area of each species type is then calculated in the following manner:

$$\text{Area of species type} = \text{Total area} \times \frac{\text{Number of points in each type}}{\text{Total number of points}}$$

Information on volume class can also be estimated for a small area around each sample point. While this method involves less photointerpretation, it suffers from not being able to provide information on location or the distribution of species types [51].

A major component of harvesting costs is the transport of felled timber to the mills so access roads must be carefully planned. Typically, access roads are of two types: temporary roads for skidding logs to collection points and more permanent roads connecting the collection points to all-weather highways. The more permanent roads will continue to be maintained after harvesting is completed for purposes of inspection, fire control, and recreation [71].

In the absence of topographic maps, large-scale aerial photos (1:10,000–1:15,000) can be used for the layout of the access road network. Wherever possible, access roads must avoid steep terrain. While roads for skidding timber must not exceed gradients steeper than 25°, many areas have limited such gradients to 10° because of the danger of erosion. Photos can be used not only to determine gradients but to detect sources of road-building materials, suitable sites for crossing streams, and swampy areas to be avoided. The selection of forest road corridors may also be made through the use of computer techniques [72].

An integral part of the forest management process is monitoring, one purpose of which is to assess the impact of harvesting practices. Since sound forest management dictates that the amount of timber harvested should not exceed the amount of new growth, estimates of harvested timber must be derived on a regular basis. A common means for doing this has become aerial photointerpretation. Specifically, the photos are used to locate and determine the acreage of clearcut timber areas. Generally, large- to medium-scale photos are used for this purpose although over large areas it means the number of photos can be considerable. A possible alternative is Landsat digital data which have also been shown to be an effective means of detecting clearcut areas. This is particularly true of band 5 and 7 data acquired during periods of snow cover [73].

Another important function of monitoring is to detect and assess the extent of forest damage. Damage to forests may be caused by any one of a number of agents – fire, insects, disease, air pollution, storms, or even recreationists. It is generally recognized by changes in tree shape, in the loss or discoloration of foliage, or in reduced tree growth. The detection of forest damage can be accomplished readily by means of photointerpretation although a number of factors must be taken into consideration including photo scale, film type, and timing both seasonal as well as daily. In regard to photo scale it must not be too small or many subtle indications of damage may go unnoticed. Murtha has reported that 1:4000-scale photos are sufficient for counts of dead trees, but larger scales are necessary for detecting signs of stress [74]. There is general agreement that conventional color and color-infrared are the best film types for damage detection. For more specific information the reader is encouraged to refer to Ciesla [75] who has compiled a good summary of the optimum film type to be used with different insect and forest types. Finally, as for the timing of photo acquisition, McCarthy *et al.* have suggested that flights be scheduled in mid-morning or mid-afternoon so as to avoid as much atmospheric convection as possible [76]. As for seasonal timing, Ashley *et al.* found that color-infrared photography acquired during the fall was best for evaluating overall tree condition and for identifying mortality, in this case caused by spruce budworm. On the other hand, color summer photography was deemed best for evaluating current defoliation primarily because the clipped, brown needles were still being held within the budworms' webbing on the branches [77].

The primary reason for acquiring data on forest damage is to enable forest managers to introduce abatement procedures. Research has now shown that aerial photos are capable of assessing the effectiveness of abatement operations. In one recently reported experiment, medium-scale (1:15,000) color-infrared photos were used to classify defoliation in both treated and untreated ponderosa pine stands. The photos showed clearly that stands where the insect population was reduced by insecticide treatment suffered less foliage damage than those not treated [78].

To be useful for management purposes data acquired by remote sensing techniques must typically be digitized and input to computerized files where they are merged with other forest data. Depending then upon the capabilities of the system and management's needs, inventory statistics may be generated by political unit (township or county), management unit, or by some other parameter such as accessibility or productivity. As for the format of the output it may be color-displayed on a graphics screen or produced as hardcopy from a printer. The advantages of computerized inventories of this type are threefold: first, there is the ease of updating the files once the base files have been created;

second, there is the speed with which various types of inventory statistics and maps can be generated; and third, the costs of both updating and generating these products are relatively low.

While the preceding comments refer primarily to commercial forestry, they also have relevance to our 'urban forests', those trees found in city parks, along residential streets, and on empty lots. Until recently, this urban tree cover was not given much consideration, but with the loss of increasing numbers of trees due to disease and air pollution concern for the remaining trees has begun to grow.

The benefits of trees to urban dwellers are considerable. They are first of all aesthetically pleasing, but they also act to absorb sound waves, moderate the climate, and trap particulate matter in the atmosphere that we might otherwise breathe [79]. While there is no single authority responsible for protecting the tree cover within metropolitan areas, many individual communities have begun to do so in the wake of the loss of so many American elms. At one extreme is Washington, D.C. which has established a computerized inventory of every tree in the District complete with life history. Although few have reached this point, many jurisdictions have begun to institute regular aerial surveys using color-infrared photography as a means of monitoring tree health. With the cost of removing a single elm tree approaching $500, communities can afford to make aerial surveys a part of the monitoring process. Such surveys become even more cost-effective when the use of the photos by other municipal agencies is taken into consideration.

9.4 Sensitive or Fragile Environments

Certain land resources are so great in extent or so apparently resistant to abuse that they are relatively safe from destruction. Other areas possess fragile ecosystems that can be destroyed quickly before their value is even recognized. It is important therefore that communities identify such environments and protect them before it is too late.

9.4.2. *Wetlands*. Wetlands represent a small but important part of the natural environment. They function as a means of recharging and purifying underground water supplies, of controlling floods and erosion, and of providing habitats for many unique forms of life [80]. Unfortunately, wetlands are also fragile environments that can be altered easily through dredging and filling. Many wetlands have disappeared over the years to residential and industrial development. Florida alone has lost over 60,000 acres of wetlands in this manner [23].

To halt the destruction of coastal wetlands, Congress passed the Coastal Zone Management Act in 1972. The bill provides funds to states for the preparation and implementation of land use management programs within their coastal areas.

While most coastal states have applied for funds, many had enacted legislation protecting wetlands well before 1972. Massachusetts, one of the earliest states to do so, passed the Jones Act protecting coastal wetlands in 1963 (later replaced by the Coastal Wetlands Act of 1965) and the Hatch Act protecting inland wetlands in 1965 (later replaced by the Inland Wetlands Act of 1968).

The management of wetlands can be broken down into three general tasks: definition, mapping, and monitoring. Definition is important for legal and administrative reasons and is therefore a prerequisite for an effective management program. While there is no universally accepted definition, wetlands may be described as moist areas subjected to periodic inundation by river stages, tidal flow, or fluctuations in the water table [47] and supporting 'a prevalence of vegetative or aquatic life that requires saturated or seasonally saturated soil conditions for growth and reproduction' [81]. This definition covers bogs, marshes, swamps, wet meadows, flood plains, mudflats and other low areas subject to tidal action. The U.S. Fish and Wildlife Service has developed a national wetlands classification system for use with its current National Wetlands Inventory but it is also meant to provide the basis for integrating wetlands data acquired at the national, regional, and state levels [82].

Mapping involves the identification and delineation of wetlands. Experience has shown that mapping wetlands from aerial photos is easier and less expensive than mapping them on the ground although on-site inspection is usually required to ascertain the exact locations of various wetland boundaries. Typically, the identification and delineation of wetlands from photos focuses on vegetation patterns, in part because they are good indicators of soils and drainage but more importantly because major plant species have distinct reflectance characteristics, particularly in the near-infrared portion of the spectrum [83]. These characteristics or 'spectral signatures' reflect morphological and anatomical differences although to obtain them species must also have high local abundance and must occur in the uppermost stratum of a stand. Because fresh-water species do not meet these criteria as well as salt-water species, they tend to be more difficult to map [84].

The timing of an overflight can be critical to species discrimination. If money were no object it would be useful to acquire photos at several points during the growing season since plant species tend to reflect differently in the near-infrared portion of the spectrum at various phenological stages. However, since money typically is a limiting factor one alternative is to acquire photos late in the growing season – in August or September. Seher and Tueller found that plant species could be identified and mapped as readily from late summer photos as from multiseasonal photos [85]. Civco *et al.*, on the other hand, suggested that it is best to acquire photos during the wettest part of the leafless season. By late summer, at least wooded wetlands would have become obscured by tree canopies [86].

All types of aerial photos provide wetlands information that is difficult to obtain on the ground, but color-infrared photos provide more detailed information on vegetation than other types. For one thing color-infrared photos are much sharper due to the film's natural haze-penetration capability. Then, too, the pinks and reds displayed on color-infrared photos have a greater tonal range than the subtle shades of green on color photos [88]. Color-infrared also makes possible the mapping of water courses and moist areas. Because water is highly absorptive of infrared wavelengths, water bodies and moist areas appear blue to black which contrasts sharply with the pinks and reds of natural vegetation. As for the ability of color-infrared to provide information on submerged vegetation, there is some disagreement. Austin and Adams concluded that color-infrared could not be recommended for this purpose [89]. Sharik *et al.* found that submerged vegetation could be easily detected. The latter suggested that an important variable is the turbidity of the water [84].

The accuracy of species identification and delineation depends heavily upon photo scale; the larger the scale employed the more accurate the identification and the sharper the boundaries (Fig. 55). The wetlands of New Jersey and New York were mapped from 1 : 12,000-scale color and color-infrared photos with the final data transferred to a 1 : 24,000-scale photo basemap. In practice the 1 : 12,000-scale contact prints were found to be too small for accurate species identification and notation in the field so 1 : 6,000-scale enlargements were used. Unless wetlands maps are to be produced for the information of landowners or for regulatory permitting procedures 1 : 24,000-scale maps should be adequate for most purposes [82].

Over large areas 1 : 100,000 is becoming a common scale for wetlands mapping. The Fish and Wildlife Service is conducting its National Wetlands Inventory at 1 : 100,000. This was also the scale at which USGS mapped the Great Dismal Swamp [90]. For this latter project several sets of color-infrared photo of different dates and scales were the primary data source with most of the interpretation done from 1 : 130,000-scale photos. Mapping at a 9-hectare (22 acres) minimum parcel size, 43 separate canopy categories and 23 vegetative communities were identified. For the final 1 : 100,000-scale mosaic base these categories were aggregated into ten canopy classes, three understory classes, and three altered vegetation classes. Field testing showed the interpretation or the original categories from the photos to be over 90 percent correct.

At this time, there are no operational programs dependent upon Landsat for wetlands data. Although a number of investigators have evaluated Landsat for this purpose it is generally recognized that Landsat lacks the necessary spatial, spectral and radiometric resolution. Management requirements and legal considerations require precise boundary placement which is best represented on maps of 1 : 24,000 or larger [82].

The final stage of any wetlands management program is monitoring, the

Fig. 55. Estaurine wetlands in the vicinity of Portland, Maine; scale 1:12,000. Courtesy James Sewall Company.

purpose of which is threefold: first, to detect illegal activities taking place within wetlands areas such as dredging and filling; second, to ascertain that permissible activities are not having a negative impact upon the environment; and third, to determine whether in light of the first two functions the regulations are proving adequate for wetlands protection. While all three of these tasks are aided by aerial surveillance, conventional aerial photos need not necessarily be acquired. Surveillance may be accomplished visually or with the use of 35 mm or 70 mm oblique photos. The advantage of aerial photos to the monitoring process is their ability to provide a permanent record of changes in wetlands that occur naturally over time. Photos may also provide before and after evidence needed for litigation when violators of wetlands regulations are taken to court.

9.4.3. *Habitat Assessment and Mapping*. The continuing conversion of open land to urban development, various government policies, and changes in farming methods have all contributed to the destruction of many wildlife habitat areas. In turn, this has led to a decline in the numbers of many species and to the near extinction of more than just a few. Congress has responded by enacting the Endangered Species Act (1973) which makes it unlawful to destroy the habitat of any endangered species. The snaildarter is the most celebrated species to be protected by this act. But the states, too, have responded and for many the rationale is financial. Game birds and animals are responsible for generating millions of dollars in revenues from hunting and fishing licenses and from taxes on food and lodging. To protect these resources, states have enacted programs for maintaining minimum acreages and a proper distribution of habitats for certain wildlife species [91].

Wildlife habitat assessment and mapping is similar to any type of land cover mapping. The actual cover types to be mapped and therefore the scale and film type to be recommended are dependent upon the region and the particular wildlife species of interest. Wetlands, one of the more important wildlife habitats, were discussed in the preceding section. Mroczynski and Eisenhauer have described how 1 : 24,000-scale black-and-white photos were successfully used to identify and map winter habitat for the ring-necked pheasant in Benton County, Indiana [92]. A decline in pheasant numbers had led the state to introduce a program for leasing land from farmers for pheasant breeding habitat. Potential lease sites would be selected on the basis of their proximity to permanent habitat. To this end pheasant habitat was delineated on the black-and-white photos and acreages were computed by means of a dot grid. Habitat maps were prepared for each of 15 townships or for a total area of approximately 2,600 square kilometers. The time required for the completion of the entire project was two months.

A number of researchers have developed computerized models for assessing wildlife habitat based upon remote sensing data. Mead *et al.* have reported on a system for assessing wildlife habitat from vegetation maps [93]. The maps were of the Great Dismal Swamp and had been produced by Gammon and Carter from color-infrared photography [90]. Vegetation data were digitized and used to derive a habitat diversity map for a hypothetical wildlife species. Criteria employed in determining habitat desirability were species preference for certain vegetation distribution patterns and vegetation/land cover edges. The latter, where types of cover and food come together, are critical for the existence of specific wildlife species. Additional factors known to encourage or restrict wildlife species (i.e., the presence of water) were also considered in the system.

Landsat data are frequently used in habitat assessment models because they are already in digital form. Lyon used them to determine the suitability of an area as a kestrel nesting habitat [94]. His objective was not to estimate numbers but to assess the potential of the area in question to attract and/or maintain a population of kestrels. Based upon the preference of kestrels for certain types of land cover, a model was developed using Landsat-derived land cover data and the spatial distribution of land cover types. Areas selected by the model as suitable kestrel habitats were fieldchecked and in 70 percent of the areas selected the presence of nesting kestrels was confirmed. Wiersema used both Landsat digital data and Landsat imagery to conduct a habitat study of the wild ibex in the national parks of the French and Italian Alps [95]. A variety of products were used to map vegetation types, detect wintering areas, and identify snow-free connections for possible mid-winter migration routes. The author found false-color composites to be best for discriminating between vegetation types and for detecting wintering areas, while band 5 data were best for analyzing snow cover.

In addition to habitat assessment, information on species numbers may also be required both for determining habitat need and for establishing harvesting regulations. Generally, animal habitats are in inaccessible locations so the acquisition of population estimates by ground survey methods poses serious problems [96]. There is the problem of finding the animals, of counting the animals, and of determining what proportion of the animal population was actually observed. But most importantly ground survey methods are extremely tedious. For these various reasons remote sensing techniques are increasingly being employed and at least for certain species are providing estimates as accurate as those generated by ground surveys. Typically, the most accurate estimates are those for large animals inhabiting open country: caribou, antelope, musk ox, elephants. A variety of low-altitude techniques have been used including direct counting, 35 mm oblique photos, and black-and-white vertical prints. All of these techniques work best if employed at times of the year when species are banding together for migration purposes [13].

A considerable amount of research on the censusing of game animals has involved thermal infrared data (emitted heat). The general conclusion resulting from this work appears to be that the thermal contrast between, for example, deer and surrounding vegetation is not sufficient for accurate discrimination [97]. More recent research suggests that multispectral data acquired during winter when animals gather into their wintering yards may provide better results [98]. Further refinement of these methods is needed.

9.4.4. *Sites of Historical or Archeological Significance*. Among the most fragile sites are those of historical or archeological significance. These are sites whose disturbance could eliminate important aspects of a local community's or a nation's cultural heritage. While many such sites are vast, like the Gran Chaco ruins of the Anasazi Indian civilization in New Mexico, the majority are small and the evidence of their existence ephemeral. The role of remote sensing in the preservation of these latter sites in particular may be critical. Aerial photos and the maps made from them may be the only way the evidence for the existence of these sites can be recorded and brought to the attention of planners and the public [99].

The use of aerial photography for archeological research dates from the 1920's although isolated examples of its use can be found before this period. In 1906 balloon photography of Stonehenge gave researchers their first view of the entire site [100]. Yet, in spite of this seemingly long record of use, remote sensing techniques are still not widely used by archeologists.

There are three areas in which remote sensing techniques can be applied to archeology: site discovery, site analysis, and site recording.

Site discovery is based on the premise that any significant cultural activity must have altered the landscape in some way and these alterations may remain visible for many years. The key is to know what you are looking for. Strandberg has described the task as similar to tactical military photointerpretation where the search is for camouflaged installations. In the case of historic sites the camouflage is nature – 'emplaced over a span of many – sometimes hundreds or even thousands – of years' [101].

The search for historic sites typically begins with a consideration of the factors influencing site selection in the past – the availability of fresh water, level land for agriculture, an easily defendable site, proximity to game habitats. Areas where several of these criteria are met are the most logical places to search for signs of early settlement. One note of caution – the physical environment of today may have been drastically altered from what it was in the past. Once open lands may now be heavily wooded, or as is more apt to be the case, once wooded areas may now be nearly denuded of vegetation.

Many early sites today lie buried under several feet of earth but their subtle outlines may be detected through plant-growth anomalies, soil variations, and shadows. Often these outlines are visible only from the air and even then only at certain times of the day or year or under very specific climatic conditions. Plant growth anomalies are caused by features such as old walls or building foundations lying just beneath the soil surface. They are best revealed in fields of a uniform cover crop and during dry years (Fig. 56). Wet years tend to eliminate crop differences [102]. Soil variations reflecting the filling in with a different material of ancient ditches or moats or cellar holes are most visible in freshly plowed fields. Shadows may reflect either plant-growth anomalies or features jutting through the surface. They are revealed best on oblique photos with low sun angles. Sun direction is also important; features may be invisible if they trend in a direction parallel to the direction of the sun.

While aerial photos for archeological purposes have been used more widely in arid regions and where a high percentage of the land is in cover crops, they have application as well in heavily wooded areas. Indicators of early settlements in wooded areas include fence lines or stonewalls, cellar holes, old clearings, the existence of ornamental shrubs, and abrupt changes in species composition [103]. As part of an Environmental Impace Statement, Ashley *et al.* used this approach in searching for early white settlement homesteads over an 80,000-acre site in northern Maine to be flooded by a proposed Corps of Engineers hydroelectric project [104]. Employing 1:20,000-scale black-and-white photos, high

Fig. 56. Roman fort, Glenlochar, England. July 1949. In years of normal rainfall little can be seen of this Roman fort, but in the drought year of 1949 the fort's outline became clearly visible. Most prominent is the street layout and the series of three ditches that surrounded the fort. British Crown Copyright Reserved.

potential sites were selected for subsequent checking in the field. In general, fencelines, stonewalls, and cellar holes are detected best on black-and-white photos acquired during the leafless season; vegetation indicators require photos taken during the latter half of the growing season, preferably in color or color-infrared.

Site analysis involves using photos to determine the extent, orientation, and significance or function of an early site. Some sites are so vast that aerial photos provide the only way of viewing them in their entirety. The arrangement of buildings, the road network, and the relationship of the site to its surrounding environment can all be better understood in this way. Photos are now being used to study entire cultural landscapes. In England this involves attempts to reconstruct whole segments of the ancient Roman landscape. One valuable byproduct of the process of reconstructing ancient landscapes is a greater understanding of present landscapes.

The vulnerability of historic sites to destruction through development, or even from excavation for research purposes, requires that permanent records be made. Accordingly photogrammetric techniques may be employed to produce detailed planimetric or topographic site maps. In the case of excavation, maps can be produced at regular intervals. These maps and the photos from which they were made will show not only what artifacts have been found but where they were located in relation to one another. In this manner, a wealth of information is stored easily and permanently.

All types of photographic emulsions have been employed in archeological research and anomalies have been detected on each. Black-and-white film has the advantage of being the least expensive and of providing high quality prints that can be taken into the field; color-infrared film is best for analyzing vegetation patterns. It has been suggested that optimally both black-and-white and color-infrared films should be used, but if only one were to be used, color-infrared would be recommended [105]. One further recommendation would be to acquire both verticals and obliques whenever possible.

Finally, there appears to be no single period of the year best suited to all forms of archeological research [106]. In fact, since evidence is so ephemeral photos must be acquired continuously. For cost reasons, then, light aircraft will probably be necessary [99] although many other types of platforms have been used. Johnson and Kase, for example, have reported that for an archaeological project in Greece, photos acquired from an unmanned, tethered balloon were used and proved less costly than conventional aerial photos [107].

10. Site selection issues*

10.1 Introduction

The preservation of land resources is a major component of the land management process but just as important is the selection of sites for development. The litany of environmental disruption caused by the improper siting of residential complexes and industrial facilities is a long one and needs no repeating here. But an important objective of land planning is to keep such disruption to a minimum. To this end, this chapter will focus on three aspects of the land development process – site selection (primarily for large public works projects), zoning and subdivision regulations, and litigation resulting from the construction of certain facilities. Specifically, the chapter will attempt to demonstrate how remote sensing techniques can contribute to these activities.

10.2 Site Selection

The process of selecting a site for the construction of a large industrial facility or public works has a number of objectives, among them to minimize the costs of developing the site, to minimize the disruption on the physical and social environments, and to minimize any potential risks associated with the proposed facility. In order to minimize site development costs, a number of engineering and construction problems must be addressed early in the site selection process. While the specific considerations tend to differ with the type of facility proposed, general concerns typically include the following: What will be the extent and difficulty of earth removal and grading operations? What is the height of the water table, and during excavation will water flow into the excavation from the water table? What is the capability of the site for providing most of the necessary construction materials? Will soils compact properly, and will they

*The material in this chapter is reproduced with permission from *Remote Sensing for Planners*, copyright 1979, Center for Urban Policy Research, Rutgers – The State University of New Jersey.

be capable of supporting the proposed facility? These are but a few of the questions for which answers must be provided. For a more complete discussion of the site development process, the reader should refer to Douglas Way's *Terrain Analysis: A Guide to Site Selection Using Aerial Photographic Interpretation* [108].

The minimization of environmental and social disruption has now become a formal process, at least for federal projects, with passage of the National Environmental Policy Act (NEPA) in 1970. NEPA requires that environmental impact statements must be drawn up for all proposed federal facilities. These documents must identify all potential environmental and social impacts and must be made available to decision makers as well as to the general public [109]. Since passage of NEPA the majority of states in the United States have adopted similar legislation. The primary objective of all such legislation is to ensure that environmental and social factors are given adequate consideration during the project formulation stage [110].

For most facilities there is little or no serious risk posed by their construction. But there are a few facilities, nuclear power plants, for example, and certain types of chemical plants, where at least the perceived risk is very great. For such facilities special consideration must be given to the nature of the risk and to the methods to be employed for its minimization. Once again, consideration of these matters must take place at the project formulation stage.

A general model for illustrating how remote sensing, and in particular aerial photointerpretation, can be used in the site selection process has been suggested by Wobber and Martin [111]. Although presented specifically for the siting of nuclear power plants, the model has applicability to the siting of all large facilities. The authors view the siting process as composed of three stages, each of which requires a certain level of information.

At Stage I likely areas are identified within which possible sites for the proposed facility may be found. Areas that are unsuitable, or at best marginally suitable, are eliminated from consideration. At this stage in the site selection process the factors examined include topography, drainage, vegetative cover, and existing cultural features; detailed information on these factors is not critical. Accordingly, any small-scale imagery is sufficient for this part of the analysis. High-altitude aerial photography would generally be recommended, but Landsat imagery (1:250,000) and even index mosaics could also be used.

Stage II has as its objective the identification of a number of potential sites for more thorough examination and from which a set of candidate sites are selected. For this, information is needed on soils, slope, geology, the presence or absence of hazards, and the location of various cultural features. Much of this information can be obtained from medium- to large-scale aerial photos (1:30,000–1:10,000).

At Stage III in the process a single site must be selected from the several candidate sites. Extremely detailed information is needed at this stage and although much of it can potentially be acquired from aerial photos, very large scales are required (1 : 1,000 and larger). Because of the large scale of the photography, the cost-effectiveness of aerial photointerpretation techniques decreases dramatically in comparison with ground surveys. As a rule, aerial photointerpretation techniques are most useful at Stages I and II; detailed ground investigation assume greater importance at Stage III. In the following pages the site selecting process will be examined as it applies to several different types of facilities.

10.2.2. *Nuclear Power Plants*. There is presently no facility for which selecting a site is as controversial as that for a nuclear reactor. Since the accident at the Three Mile Island nuclear facility in 1979, increasing concern has been shown for plant safety. In at least one case, the Diablo Canyon nuclear power plant in California, anti-nuclear forces held up final licensing of the completed plant for several years. While the United States Nuclear Regulatory Commission (NRC) has responded by enacting stiffer construction and operation guidelines, there is now serious doubt as to whether the construction of any new plants will ever be undertaken. Nevertheless, the process of siting a nuclear facility illustrates well how remote sensing techniques can contribute to the overall site selection process.

Experience has shown that while many different companies have been involved in the siting and construction of nuclear power plants, there has been considerable similarity in the way each has approached the site selection process. At the preliminary stage in the siting process (comparable to Stage I in the Wobber/Martin model) two concerns tend to be paramount – identifying sufficient sources of water for cooling purposes and the need to minimize risks to surrounding populations. Areas lacking sufficient water and/or containing heavy population densities are typically rejected as candidate areas. The most ideal areas are sparsely settled farming regions or marshlands located adjacent to large rivers or the ocean. Since the principal purpose of this stage is to exclude unlikely areas, almost any small- to medium-scale imagery is capable of providing the necessary information, although it is uncertain how many companies have employed remote sensing techniques for this purpose.

The selection of several potential sites from within the candidate areas (Stage II) would require the use of larger scale photos. Of particular significance at this stage is the identification of any features that may add to the cost or the risk of developing a particular site. For example, a major component of construction costs is the cooling system. Therefore, each potential site must be capable of providing sufficient water for cooling purposes throughout the year including periods of low stream flow. If the water supply proves unreliable a reservoir may have to be built adding to overall construction costs. Much the same can be said for the risk factor. Care must be taken to ascertain the existence of any geological hazards in the vicinity of potential sites – faults, fractures, areas of subsidence or areas prone to flooding. Only after they were completed were faults discovered running through nuclear reactor sites at Lake Anna, Virginia and Wiscasset, Maine. Had aerial photos been used in selecting the sites for these plants, the faults may well have been discovered before construction ever began.

Geologic information is not the only information needed at this stage; detailed data are needed on the distribution of population and on various cultural features. Reactor accident calculations suggest that risk is a function of both the total population surrounding a site as well as the number of people within any given direction sector [112]. Population estimates can be derived from the photos (see Chapter 11). Particular attention must also be given to the transportation network. Not only must there be sufficient access roads in terms of number to handle the traffic generated by the construction activities, but they must be free of steep grades because of the extreme weight of the reactor components.

Much of the information required at Stage II can be acquired from good quality black-and-white stereo prints such as those available from USDA. But it would also be useful to supplement the black-and-white prints with a good set of color-infrared transparencies. These transparencies could be a source of data on drainage and soils, and particularly on vegetation patterns which could be a surrogate data source for soils and drainage. Furthermore, land use information can be more easily acquired from the color-infrared transparencies than from black-and-white because of the sharp contrast in scene elements. While the cost of obtaining both might be too great for the siting of some facilities, it would appear to be an almost negligible expense in the case of nuclear power plants that measure their construction costs in hundreds of millions of dollars.

Final selection of a site at Stage III requires very large-scale photography if aerial photography is to make any contribution at all. Though typically the contribution is small in quantitative terms, it may be important nevertheless. For example, remote sensing techniques were used successfully to evaluate an area of karst topography as a potential site for an industrial development [113]. It was found that color-infrared could reveal up to 85 percent of the sinkholes and locations of solution activity. Such information is important because it influences the layout of a facility and the design of the foundation. Aerial photos will be employed in this manner as long as their use and contribution can provide some savings in time or money.

Most of the data employed in siting nuclear power plants and other facilities for that matter can be digitized and input to geographic information systems for computer analysis. The use of computers in this way makes possible the application of larger data sets and various weighting techniques; it also provides for the visual display of results. Hansen has reported on how such a technique was employed to assess the effect of population density and certain environmental factors on the availability of land within the coterminous 48 states for siting nuclear power plants. The study conducted by Dames and Moore for the NRC had as its main objective the discrimination of more suitable siting areas from less suitable areas [112].

To conduct the analysis, the 48 states were subdivided into 5-km square grid cells totalling 600,000 in all. Of the various data input to the system population data were the most detailed and had the further advantage of already being available in digital form. Data on urban centers and population density were employed in the analysis. Several types of environmental data were acquired – on slope, water availability, and seismic probability. For each of the environmental data sets a suitability map was produced using a weighted scale (1 least suitable, 9 most suitable). Additional information was added on restricted land, that is, land legally protected and therefore unavailable for development. A composite suitability map was then produced from the four map files with individual cell values ranging from 4 to 25. To identify the most suitable sites various types of population data were overlayed to the environmental suitability data (e.g., adequate water supply, low seismicity, gentle topography, and an absence of restricted land). The most suitable sites correlated strongly with areas of high population density. There also tended to be more suitable sites in the eastern half of the country than the western half.

The study did not employ land use data *per se* but as Wray has pointed out, land use and land cover maps produced by USGS are expected to be available for the entire country by 1986 [114]. Soon after those same data will be available in digital form. The easy availability of such data will further hasten the use of computers in the site selection process.

10.2.3. *Solid Waste Disposal.* One of the many by-products of industrial growth is the generation of ever-increasing quantities of solid waste. Historically these wastes have been burned in open dumps but legislation aimed at improving air quality now prohibits this method of disposal. The alternatives available are several – sanitary landfills, recycling, incineration, or a combination of these methods. Local communities faced with choosing among these alternatives may find that remote sensing techniques can contribute to the decisionmaking process.

The selection of an appropriate waste disposal system depends in large part on the amount of waste being generated. A technique for estimating solid waste generation from aerial photos has been suggested by Garfalo and Wobber [115]. The technique involves the use of color-infrared photos to enumerate and classify buildings by type. This was done for the city of Tampa, Florida. Next, the mean number of residents or employers were estimated for each building on the basis of average occupancy data gathered by the United States Bureau of Census for Tampa. Finally, these figures were multiplied by national averages of waste generated per capita per day (derived by HEW) to produce estimates of solid waste generation for districts within the city and for the city as a whole. The authors claimed that the estimates generated in this way were consistent with those derived by other means but were generated more quickly and less expensively.

Once a method of waste disposal has been determined, the process of selecting a site for the facility may begin. Sanitary landfills are the most commonly selected methods because of their simplicity and capacity to handle all kinds of wastes. A landfill site is prepared by digging a trench or cutting into a slope with some type of earthmoving equipment. Waste materials are dumped into the excavation and covered with earth on a daily basis. If sited and operated correctly there should be no blowing papers, no odors, no fires, and no infiltration of ground waters by contaminants from the waste.

The most ideal site for a landfill operation is a natural depression underlain by an impervious stratus; the latter will prohibit contaminants from leaking into the ground water [108]. In addition, the water table should be several feet below the bottom of the depression to further reduce the risk of contaminating ground water supplies. There should also be sufficient materials to cover the waste. These materials would ideally be of an impervious nature to keep rain water from penetrating into the waste and creating a contaminated substance or leachate. Unfortunately, these conditions are seldom met and all too often ground waters become contaminated by the disposal procedures.

For the selection of potential sanitary landfill sites, Stages I and II of the Wobber/Martin model can be combined through the use of medium-scale photography. Potential sites should be located away from residential areas, yet they must be connected to a highway or road network capable of handling the traffic generated by the facility. Sites should also be located away from local water supplies. Here color-infrared photography would be particularly useful since it clearly displays the drainage network and areas where the water table is close to the surface. Finally, potential sites should be large enough to accommodate the projected quantities of solid waste and should contain sufficient parent materials to make possible the regular burying of disposed materials.

If a waste disposal facility is to serve a number of communities, it may be necessary to select several sites where waste materials may be collected and either recycled or compacted before being transferred to the disposal facility. Aerial photos or orthophoto maps can readily supply the information needed for selecting these sites. The latter should be conveniently located and served by a highway network sufficient to handle both a large number of users and the heavy trucks required to transfer the concentrated waste to the disposal facility. The traffic generated by collection sites mandates they be located at the periphery of built-up areas.

10.2.4. *Flood-Control Dams*. Like nuclear power plants, the construction of flood-control dams has aroused much concern. At issue are the vast areas behind the dams that are periodically flooded. In recent years as opposition to such dams has grown, the United States Army Corps of Engineers, the agency primarily responsible for our nation's waterways, has increasingly turned to non-structural means (e.g., floodplain zoning) to minimize flood damage. Nevertheless in those instances where structural controls provide the only alternative, aerial photointerpretation techniques can provide much of the information required for the site selection process.

Medium-scale photos can provide the information needed to minimize the environmental and social disruption caused by the construction of flood-control dams. During periods of rapid runoff such as in early spring when mountain snows are melting, sluice gates may be closed to reduce the flow of water downstream. The effect is to flood vast tracts of land upstream from the dam. Often, as in New England, these flooded tracts are among the region's best agricultural lands. Aerial photos can be used to estimate for each potential site the number of acres to be flooded and the current use of that land. Good stereo coverage is needed. The photos can also reveal the number and types of buildings that may have to be removed and the roads that may have to be rerouted. Because the costs of acquiring land and moving households can be very great, particularly if litigation results, the estimates provided by the photos can be critical to the site selection process.

Additional information characteristic of Stage III will require large-scale photos. Determination must be made as to whether potential sites possess geologic formations capable of supporting the flood-control structure. Photos can also be employed to estimate the quantity of sediment that would be carried downstream in the event of flooding [47]. Depending upon the extent of the flooding, a portion of the floodplain's surface soils will be swept away and deposited behind the dam. Over a period of time these deposits can build up and render the dam less effective. The areal extent of floodplains can be determined by aerial photointerpretation techniques though depths typically have to be calculated from ground measurements.

10.2.5. *Corridor Selection.* Corridor selection refers to the process of selecting a route for a new highway or transmission line. The process is similar to site selection – it shares the same objectives and typically proceeds through a series of stages. Higway organizations have certainly been the most frequent users of the corridor selection process and have done the most to refine it. Historically, earthwork and engineering considerations received primary emphasis in highway planning but since passage of the National Environmental Act (1970) the corridor selection process has changed dramatically. Whereas the 'best' route for a highway in the past was the one producing the greatest savings in cost or travel time, today it is more apt to be the one most effectively reconciling the conflicting interests of a host of groups affected by the proposed route [116]. This change in emphasis has necessitated that an increasing number of social and environmental factors be considered in the selection process. In turn, this has led to an increasing reliance on computers to aid in highway route selection.

Aerial photos have been used in highway planning for many years and their importance has increased, particularly as social and environmental concerns have become a more important part of the route selection process. The literature on this topic is extensive and while the new *Manual of Remote Sensing* (1983) probably contains the most complete and up-to-date set of references, a source deserving special attention is Hamilton and Locate's *Using Air Photo Interpretation and a Socio-Ecological Reconnaissance in the Highway Route Selection Process* [117].

This same concern for social and environmental factors has also led to the use of aerial photos in the routing of powerlines. Klunder and Arend have described the process for selecting a right-of-way for a 52-mile powerline (450 KV) through northeastern Vermont [118]. In spite of the rural nature of the region, the constraints imposed on those selecting the route were considerable. First, the State of Vermont has restrictions on routing powerlines through highly developed farming areas. Research shows land is lost to production where this occurs [119]. Then, there are health and safety concerns that have been raised about the presence of such lines. While the public's perception of these issues may not always be a valid one, power companies make every effort to avoid concentrations of population. And finally the State of Vermont has a comprehensive set of environmental protection laws covering archeologic sites, wildlife habitat, and unique scenic areas. The combined effect of these constraints is to make the route selection process complex but also to make the use of aerial photointerpretation techniques more necessary.

The route selection process described by Klunder and Arend was identical to the model presented by Wobber and Martin. Stage I was conducted with the aid of existing 1 : 84,000-scale color-infrared photos that had been enlarged to 1 : 42,000. These photos allowed for land use and land cover mapping over the entire study area. In addition, terrain characteristics, drainage, wetlands, the transportation network, and population distribution were also mapped from the photos, each as a separate overlay. This series of overlays was used to produce a final map which indicated areas of no limitations, slight limitations, and moderate to severe limitations to the construction of a powerline. On the basis of the final map a series of 29 segments, one mile wide and varying in length from 2 to 20 miles, were selected and interconnected to show several possible routes for the powerline.

For Stage II in the route selection process larger scale (1 : 24,000) black-and-white photos were acquired of the various segments; 1 : 6,000 enlargements were also obtained. These photos were used to identify and map the vegetative cover, individual structures, public and private roads, existing utility lines, wetlands, and fragile geological areas. These were generally mapped at about one-acre minimum parcel size. Then, where possible, property lines were added to the maps as well as property tax information. The end result of this stage was the selection of a preferred 1,500-foot right-of-way within which the powerline would be built. The preferred route avoided, as much as possible, conflict with existing land uses, year-round and seasonal homes, and natural constraints such as steep terrain and exposed bedrock (Fig. 57).

The final stage (Stage III) involved establishing the exact route of the line. For this purpose detailed information was needed on topography, the location of access roads, and on the various environmental constraints – wetlands, soil and subsoil conditions, and vegetation patterns. These data were gathered in part from aerial photos but had to be significantly added to by field investigations and helicopter overflights. Perhaps most importantly, the route selected in this manner was accepted by the State of Vermont with only one minor revision, and that based on aesthetic criteria.

10.3 Zoning and Subdivision Regulations

Zoning and subdivision regulations are two of the most significant tools available to local governments for regulating private land and building development within their jurisdictions. Zoning ordinances stipulate the land uses permitted in various districts of a city or town as well as other restrictions that may apply to those districts – minimum parcel sizes, maximum building heights, and possibly even architectural limitations. The purpose of subdivision regulations is to control the arrangement of structures and roads in new developments so that they fit the environment of which they are a part. Both types of regulations may potentially take advantage of aerial photointerpretation techniques.

Zoning ordinances permit land owners to request a rezoning or zone variance if the owner feels the existing zoning is inappropriate or unfair. The request is usually made to the zoning board or a special board of adjustment established specifically for the purpose of hearing such petitions. In deciding whether to authorize a variance board members must determine how the proposed change will affect the neighborhood as a whole and adjacent properties in particular. Relevant information includes surrounding land uses, structures, and traffic circulation. Unfortunately, many boards lack sufficient money and time to evaluate requests carefully, particularly when this means visiting each site on the ground. In such instances aerial photos or even orthophoto base maps can be of great help because they can provide much of the necessary information quickly and easily. Furthermore, the aerial view will show the relationship of the site under consideration to the surrounding environment and at the same time it will

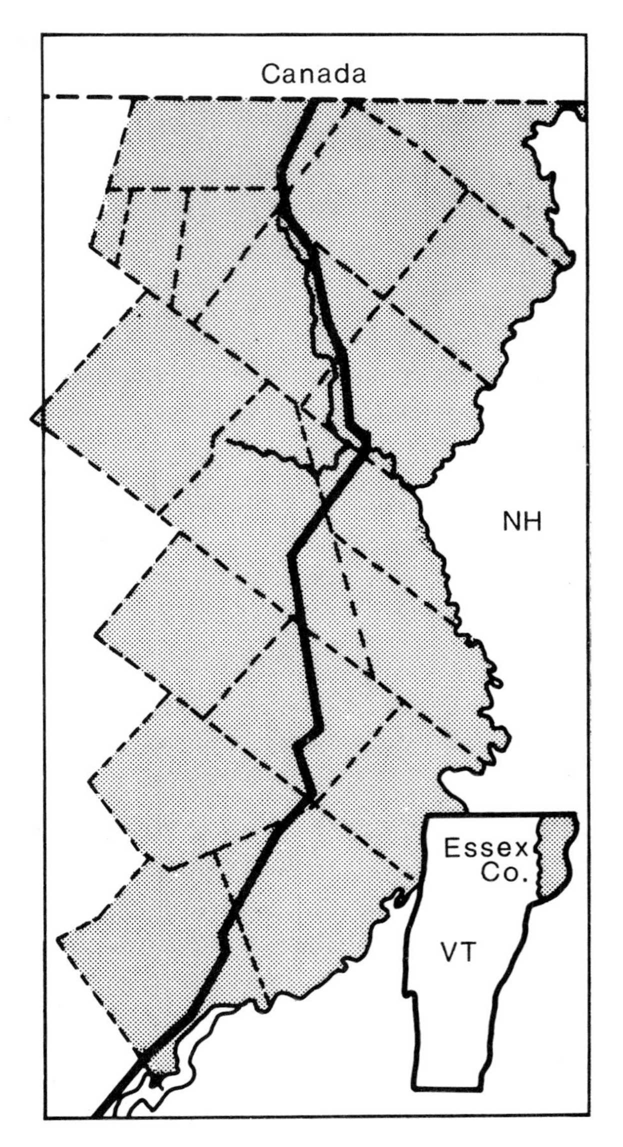

Fig. 57. Final route selected for the 52-mile VEPCO powerline through northeastern Vermont. Had the route been a straight line, it would have extended only 47 miles.

allow board members to check on the validity of information presented in favor or against the proposed change without the necessity of going back to the field. Finally, Branch has noted that there may also be legal advantages to the use of aerial photos for rezoning purposes [120]. At least some state courts, California for one, have ruled that ground surveys may violate due process unless certain procedural requirements such as advance notice of the visit or advance indications of the findings are met. More will be said about the legal applications of aerial photos in the next section.

Similarly, aerial photos can provide information relevant to the evaluation of proposed subdivisions. As a means of determining whether a new subdivision will meld into an existing environment, photos can provide information on slope, drainage, vegetation, access, utility lines, and surrounding development. Whereas for rezoning purposes photos would typically be studied monoscopically, stereoscopic analysis would be more appropriate for subdivision evaluation and monitoring.

10.4 Legal Implications of Remote Sensing

In preceding chapters frequent references have been made to the use of aerial photos and remote sensing techniques for litigation purposes. This is becoming an increasingly important function and accordingly requires further elaboration. The following discussion will focus on two facets of this issue – the use of aerial photos as adjudicatory evidence and the use of aerial photo and remote sensing techniques in statutory enforcement.

The use of aerial photos (and maps made from them) as evidence in court goes back many years, nearly in fact to the turn of the century. Yet in spite of this long history of use, the legal acceptability of aerial photos has proceeded slowly and with some difficulty. The concern over the admissibility of aerial photos has arisen in large part from the possibility of altering photos during processing. Since cases exist where 'doctored' ground photos have been entered as evidence, lawyers and judges have tended to be wary of aerial photos about which they generally know very little [121].

Judges have been willing to admit aerial photos as evidence only when alternative ground data have been unavailable and when the photos can be properly authenticated. In the past authentication has frequently required the presence of the actual person taking the photos but today the testimony of an expert witness is usually sufficient. Typically the expert witness is a professional engineer or surveyor who has familiarity with photogrammetric and aerial photointerpretation techniques. To authenticate the photos the expert witness must be prepared to answer questions about the photos such as: the name and reliability of the company taking the photos; the date of the photos; of the condition of the atmosphere at the time the photos were taken; the altitude of the aircraft; the type of camera, film, and filters used; and the way in which the photos were processed. As for maps prepared from photos, lawyers are usually concerned only about their relative accuracy; they will seldom ask questions about the photogrammetric techniques used to compile them.

A recent case involving the United State Army's Corps of Engineers provides a good example of the use and acceptability of aerial photos in the courtroom. The case involved a group of landowners who sought to recover compensation from the United States Government for damages caused to their property by the construction and operation of several navigation locks and dams on the Ohio River [122]. At the same time the owners also challenged the use of aerial photos in drawing up the legal documents describing the flowage easements taken on the plaintiffs' property by the Corps.

In response to the first allegation that the construction and operation of the locks and dams was responsible for the erosion of riverbank properties the Corps of Engineers undertook a study of the erosion process along the Ohio River both prior to and after the construction of navigation facilities. Several sets of aerial photos were employed to document changes in streambank erosion for the period from 1938 to 1977 (the navigation structures were built during the 1950's). The conclusion of the photoanalysis was that bank erosion would have occurred with or without the construction of the navigation facilities since most damage took place during flood conditions when gates were left open to let the flood waters rush through. The plaintiffs were therefore unable to recover damages.

As to the validity of using aerial photos for the preparation of legal descriptions of the flowage easements, the judge again upheld the Corps of Engineers. The plaintiffs had contended that the photogrammetric process had not been properly authenticated by expert witnesses. The judge's finding in favor of the Corps was based on the plaintiffs' inability to provide evidence that the descriptions prepared from the photos were inaccurate.

Statutory enforcement applications are not usually intended to produce evidence for litigation purposes but rather to determine whether provisions of certain laws are being violated. For this purpose aerial photos are employed as a means of augmenting the investigative resources of regulatory agencies [123]. At least one regulatory agency, the Environmental Protection Agency (EPA), maintains its own aircraft for monitoring purposes. The EPA's enforcement program is directed primarily towards detecting violations of air and water quality statutes but recently increased interest has been shown in detecting the illegal storage of chemical wastes [124].

Another area where increasing use is being made of aerial photos is in the monitoring of stripmining activities. Coal, in particular, has been stripmined since the 1890's and over the years thousands of hectares of land have been devastated by this practice. Ground scarring, erosion, spoil banks, landslides, and polluted streams are typical features of unreclaimed stripmining sites. Historically, coal companies have been able to 'scrape and run' because their considerable political influence has made it difficult for state legislatures to enact laws requiring those companies to reclaim stripped land. At least in theory this situation has changed with passage of the Surface Mining Control and

Reclamation Act (1977) which establishes minimum national standards governing the surface mining of coal. Enforcement is left to the individual states who may deny mining permits to companies violating stripmining statutes.

Title IV of the Act has established a national program to reclaim lands mined prior to August 1977. Administered by the Department of Interior's Office of Surface Mining, the program would cover unreclaimed lands and those partially reclaimed but still endangering the health and safety of the public. As a part of this program the State of Pennsylvania undertook an Abandoned Mine Land Inventory the purpose of which was to accurately identify and map all surface features and disturbances from mining sites [125]. One portion of the inventory was conducted from 1:80,000 black-and-white photos. Nearly all surface expressions of mining activities abandoned prior to 1977 were detected and accurately delineated on the photos. Quality control was maintained by field-checking. The authors of the report concluded that had the inventory been conducted entirely by field survey methods the final cost would have been at least eight times higher.

In a similar study Mroczynski and Weismiller found color-infrared photos to also be effective for identifying abandoned mine land [126]. In this instance 1:120,000-scale images enlarged to 1:30,000-scale prints were employed to identify several categories of non-reclaimed land including barren soil, gob, and slurry ponds. Identification from color-infrared proved highly successful as 98 percent of the sites were correctly classified by category.

For lands in the process of being mined a more continuous monitoring system is required. Research suggests that Landsat may be capable not only of monitoring the progress of surface mine operations but of the reclamation process as well. Spisz and Dooley [127] used Landsat digital data to classify a surface mining area in eastern Kentucky into six land cover classes representative of various stages in surface coal mining operations – two classes of undisturbed forest, two classes of barren land, and two classes of revegetated land (greater than 50 percent vegetation cover and less than 50 percent vegetation). Classification was done on Landsat data of three dates (July 1973, August 1976, and April 1978). Comparison of results revealed the gradual enlargement of newly mined areas and the reclamation of older mined areas. Forested and barren areas were identified with 90 percent accuracy but partially revegetated areas were accurately identified only 60 percent of the time. The computer had difficulty discerning areas in the early stages of reclamation from spoil banks sprouting a sparse cover of natural vegetation.

Davis *et al.* have suggested further that Landsat digital data can be used to provide information on reclamation effectiveness [128]. Under present mining laws companies are encouraged to revegetate stripped areas as rapidly as possible. If proper soil fertilization and stabilization measures are not taken first, however, many species may die within a year or two. To test this theory Landsat digital data for several dates between 1973 and 1981 were separated into six classes using an unsupervised classification algorithm: water, water (with heavy sedimentation), vegetation, reclaimed, barren, and mixed pixels. Reclaimed areas were easily separated though over the eight-year period some reclaimed areas moved into the vegetation category. Significantly, the analysis of the several data sets revealed areas which experienced a vegetation breakdown, that is, barren areas previously classified as reclaimed. Although such areas were small the authors contended they represented a good illustration of the effectiveness of Landsat in providing data on mining reclamation activities.

11. Population and housing data*

11.1 Introduction

An important input to the land management process is timely and reliable information on the number of people residing in an area. This is even more necessary when such areas are experiencing rapid population growth. In the United States the principal sources of demographic and housing data are the federal censuses but they are systematically acquired only on a decennial basis. For rapidly-growing areas these data become obsolete within a very few years. While Congress has discussed the idea of a less-detailed census to be conducted five years after each national census, no legislative action has been taken as yet. The reason is the cost. The 1980 census cost United States taxpayers over one billion dollars. For the moment then planners must continue to depend upon statistical methods of estimating population.

In the Third World, where population growth rates have reached the highest levels ever, reliable data are needed more urgently than in the United States. Unfortunately, few national censuses have been conducted and when they have the data are seldom reliable. A combination of inadequate maps, poorly trained enumerators, a highly mobile population, vast distances, civil strife, and limited financial resources has made it all but impossible to conduct the type of census common to the more industrially-developed nations. Still, they must use whatever demographic data are available.

This demand for demographic and housing data has led to the development of several techniques for estimating population from remote sensing imagery. Most of these techniques have been tested in the United States although the extent of their use here is uncertain. Where the techniques have been applied is in the Third World where alternative data are unavailable. While the literature reveals little in terms of the financial costs of using such techniques, it does suggest that remote sensing has the capability of providing timely and relatively accurate population data [129].

*The material in this chapter is reproduced with permission from *Manual of Remote Sensing*, Second Edition, copyright 1983, by the American Society of Photogrammetry.

11.2 Population Estimation Techniques

On the basis of published research, three techniques have been developed for estimating population from remote sensing imagery. As Henderson has pointed out, these techniques do not involve the direct counting of people but rather they employ various housing and land use characteristics that are visible on the imagery to infer numbers of residents [130]. The three techniques therefore are referred to by the characteristics or surrogates they employ: the dwelling-unit, the land use or area-density, and the built-up area. Each of the techniques differs in terms of the accuracy of results, the usefulness of results, and the extent to which it has been successfully applied.

11.2.2. *Dwelling-Unit Technique*. The use of dwelling units as a surrogate for estimating population has been shown to be a valid technique [131]. In its simplest form the technique can be expressed as follows: Estimated Population = Number of Dwelling Units × Average Family Size. A number of studies have demonstrated how dwelling-unit estimates can be derived from aerial photography [132; 133; 134; 135]. Using various scales of imagery ranging from 1:4,500 to 1:20,000, dwelling-unit estimates have been derived on the basis of such criteria as form and structure of roof, number of chimneys, number of stories, relative size of structure, presence of sidewalks and footpaths, presence of garages, carports, driveways and parking areas, and amount and quality of vegetation. Although results have differed slightly in terms of the degree of accuracy, the trends have been consistent. For example, numbers of residential structures have consistently been identified with great accuracy (99 percent) with a tendency to slightly underestimate. Dwelling-unit estimates have been less accurate although with the same tendency to underestimate. Typically, the accuracy of dwelling-unit estimates has decreased as the percentage of multi-family housing has increased.

The dwelling-unit method works well in the United States because so many residents are housed in single-family detached structures (Fig. 58). Clayton and Estes have shown that in suburban areas where single-family houses predominate even high-altitude aerial photos can provide accurate data [136]. House counts derived by manual photointerpretation techniques from 1:63,360-scale color-infrared imagery of Goleta Valley, California, were compared with census data and were found more accurate. Furthermore, when the two data sets were compared on a block-by-block basis the average imagery error was less than half that of the census data and displayed no apparent spatial bias. These results confirm not only that accurate estimates can be produced by this method but that it can also be used to check census figures.

For many of the same reasons the dwelling-unit approach would appear to have applicability to Third World countries, particularly to those in which populations are predominantly rural. Just such an example was reported on by

Fig. 58. Residential area in western Massachusetts. Scale 1:4000. In areas such as this both house counts and dwelling unit estimates can be made with great accuracy. Courtesy Tom Marler.

Allan and Alemayehu who applied the technique to a rural farming area in Ethiopia where population data had never been acquired [137]. With 1:20,000-scale black-and-white photos as their data source, the authors employed a plot-sampling method, commonly used in forest inventorying, to count dwellings. Seventy-three photos were involved and the principal point of each was selected as the center point for the sample plots. Four concentric circular plots, each double the area of the preceding one, were drawn about the principal point of each photo. Dwellings were counted within each sample plot and multiplied by an average household size of five. While there were no alternate data with which to compare the results, the authors were satisfied that the estimates were useful.

The effectiveness of this approach depends not only upon accurate dwelling counts but on realistic figures for average household size as well. While Allan and

Fig. 59. Dar Caid Mesguini in Ben Sergao, Morocco, is an abandoned fortified dwelling in which a number of squatters have constructed living quarters. (ITC, 1983).

Alemayehu used an approximate figure of five, a more effective technique has been suggested by Lo and Chan [138]. Using an area in the New Territories of Hong Kong as a test, Lo and Chan developed a typology of dwelling types based upon aerial photographic and field analysis. Four distinct patterns of rural dwellings were identified and average household sizes determined for each. And whereas average household sizes are usually derived by field sampling out of necessity, the authors concluded that census methods provide more precise estimates of average household sizes than field surveys and therefore should be employed whenever available.

For all the apparent advantages to the dwelling-unit approach, it does have several important drawbacks. For one thing, it is a tedious method. Estimating dwelling units over large areas is a laborious task. Fortunately for Western societies it is becoming increasingly possible to integrate housing data from other sources (building permits, utility connections) into the calculations. For Third World countries, however, such data are seldom available.

There are also problems of a photointerpretation nature. For example, residential structures may easily be confused with non-residential structures. Eyre *et al.* found for Jamaica that seasonal and temporary shelters were often mistakenly identified as permanent dwellings [139], while Lo and Chan mention the confusion of farm buildings with residential structures [138]. Vegetation can pose another problem by masking dwellings constructed specifically to take advantage of the shade. In the humid middle latitudes the problem can be

Fig. 60. A portion of Manzese, the largest squatter area in Dar es Salaam, Tanzania. Courtesy Paul Hofstee, ITC.

minimized by acquiring imagery during the leafless season, but in the tropics, there is little to be done.

A final comment is in order concerning appropriate film types. Most of the studies evaluating the dwelling-unit approach have utilized black-and-white photography. Clearly, black-and-white photos are the least expensive and over vast areas where the individual photos may number in the hundreds, cost may be an important consideration. However, under certain circumstances the use of color-infrared imagery may be justified. The natural haze-penetration capability of color-infrared film may well improve the accuracy of dwelling counts, particularly in urban areas. Also in more humid areas the color contrast between natural vegetation and residential structures may likewise improve dwelling count accuracies.

11.2.3. *Land Use/Area Density*. The land use/area density method generates estimates of population by measuring various types of residential land use and multiplying the results by the average population density for each type. Among the first to experiment with this technique were Collins and El-Beik who used it to estimate the population of Leeds, England [140]. Limiting themselves to three 'common' types of residential structures (semi-detached, terraced, and back-to-back) which they could identify with over 90 percent accuracy, the authors selected a series of enumeration districts comprised exclusively of one or the other of these house types. For the districts consisting exclusively of semi-detached structures, population estimates were derived by the dwelling-unit method. But for those districts composed of either terraced or back-to-back housing the area density method was employed. The latter involved two tasks. First, the square footage of each house type was estimated by measuring it in strip lengths from the photos. Then, the totals were divided by what the authors referred to as the photographic factor (PF) to obtain population estimates. The PF was the average population density per square foot for each house type, a figure derived from census data and Ordnance Survey maps.

To assess the validity of their estimates, Collins and El-Beik compared them to census data. However, the results of this comparison must be used with some care because of a two-year difference in age between the census data and the photos. The results showed that population estimates for districts containing exclusively terraced housing exceeded census figures by 0.8 percent, estimates for districts of back-to-back housing exceeded census figures by 0.3 percent. Considering that the photo data were the more recent, these findings are at least consistent. The results obtained for districts containing semi-detached housing and for which the dwelling-unit method was employed were more varied. On the average these latter estimates were about 6 percent lower than comparable census figures. While this is a fairly large difference it is also true that the dwelling-unit approach tends to underestimate populations.

The same technique was applied by Kraus *et al.* to four California cities in order to determine what effect city size has on the results [141]. Employing 70 mm black-and-white imagery optically enlarged to about 1:40,000, the authors mapped four categories of land use: single-family residential (R_1), multi-family residential (R_m), trailer park (T_p), and commercial/industrial (C). Population densities for the three residential land uses were derived from census block data. Population estimates for each city were calculated as follows: $P = (A_{R_1} \cdot D_{R_1}) + (A_{R_m} \cdot D_{R_m}) + (A_{T_p} \cdot D_{T_p})$. P represents the total estimated population; A_{R_1}, A_{R_m} A_{T_p} represent the areas devoted to each land use, and D_{R_1}, D_{R_m} and D_{T_p} represent the population densities for each land use.

The results, presented in Table 16, proved very encouraging. The authors concluded that had larger scale photos (approximately 1:20,000) and a larger number of land use categories been used, more accurate population estimates would probably have resulted.

Table 16. Population estimation data

	Fresno	Bakersfield	Santa Barbara	Salinas
Estimated population	235,270	170,226	142,790	54,866
Actual population	259,028	180,263	133,437	58,896
Percent error	9.17%	5.57%	7.00%	6.84%
	(–)	(–)	(+)	(–)

Taking the recommendations of Kraus, *et al.* into consideration, Adeniyi has applied the technique to Lagos, Nigeria, a city experiencing explosive growth but also one where reliable population data are virtually nonexistent [129]. The estimation procedure involved three steps: 1) the identification and classification of residential areas by manual photointerpretation techniques; 2) the collection of population data in the field as a means of generating average population density figures for each type of residential area; and 3) the calculation of population estimates. In addition, Adeniyi conducted an analysis on the socio-cultural and physical variables to determine which ones best explain the differences in population density from one area to another.

The identification and classification of residential areas was based upon stereo-interpretation of 1 : 20,000-scale black-and-white photos. Nine residential classes were identified from a number of sociocultural variables visible on the photos. Included among these variables were building type, number of stories, density of buildings, plot size, and landscaping. Having then delineated these classes a number of sample blocks were selected within each for field surveying. Data were collected on average household size and average number of households per dwelling which in turn were used to compute an average population density per hectare for each residential class. The final step involved computing the total area of each residential class (in hectares) and multiplying those totals by the appropriate average population density per hectare figures. Unfortunately, the results generated from this approach could not be rigorously assessed.

The analysis of the variables accounting for the differences in population density between classes showed density of buildings and average population per building to be the most important variables. Density of buildings, which can be determined from aerial photos, accounts for 60 to 80 percent of the variation in population density, average population per building accounts for much of the rest. In general, average population per building increases in importance as population density increases. As a result, Adeniyi recommends that extra field data should be collected in areas of high population density.

As the foregoing examples have demonstrated, the land use/area density technique is most applicable to densely populated areas where individual dwellings cannot be accurately delineated. The most extensive areas of this type are found in the Third World where unplanned, irregular, and makeshift structures have become the dominant form of urban housing. But there are additional advantages of this technique to Third World planners. For one, the approach requires less detailed photointerpretation than the dwelling-unit method, although it may involve more field checking. Then, too, the approach is less bothered by the presence of vegetation. Generally, the heaviest vegetation, in the tropics and elsewhere, tends to be concentrated in the less densely settled areas [142].

11.2.4. *Built-Up Area.* Perhaps the least useful of the three population estimation methods, the built-up area, as its name implies, is based upon the relationship between the built-up area of settlement and its population – a relationship examined by Nordbeck [143]. An early test of this relationship was undertaken by Holz, *et al.* in the Tennessee River Valley area where a file of sequential aerial photos was available for use [144]. The authors, employing step-wise linear regression, tested the relationship of four variables to population size, including built-up area, transportation links, population of nearest largest settlement, and distance of nearest largest settlement. Their results revealed marked variations between population and area for different sized settlements. They concluded that as the population of a settlement increases, variables other than the size of the built-up area become increasingly necessary for accurately estimating a settlement's population.

Ogrosky later applied this same test to cities in the Puget Sound area but using high altitude (1 : 135,000-scale) color-infrared transparencies for acquiring the necessary data [145]. Again a high degree of linear association was found between the region's population and the four variables. Contrary to Holz's *et al.* findings, size of urban area was by far the best single estimator of urban population.

With the relationship between built-up area and size of population strengthened by Ogrosky's research, Lo and Welch applied this approach to cities of Mainland China using measurements of built-up areas from 70 mm Landsat transparencies [146]. In spite of the fact that the transparencies were dated 1972–1974 and the population statistics with which the results were compared dated 1970, the approach appeared to work well. However, because of the small-scale of the Landsat imagery the technique is limited to cities of at least 500,000.

In the three examples mentioned, measurements of built-up areas were made manually. Anderson and Anderson have compared such measurements made by human interpreters with those derived by an analog-digital imaging processing system. They found that the results of both techniques were similar to those appearing in published documents [147].

11.3 Housing Quality Data

Only a few countries in the world are equipped to systematically gather and analyze housing data. In the United States, this responsibility rests with a score of federal, state, and local agencies. The methods of analyzing housing data have increased in sophistication over the years due primarily to the use of computers. The methods of acquiring data, on the other hand, have changed little and remain laborious and expensive. Although few housing agencies, even in the United States, have considered employing remote sensing techniques for acquiring housing data, research results in this area have been encouraging.

Data acquired by housing agencies are basically of two types. One is concerned primarily with the structural condition and quality of individual residential units; the other is related to the residential or neighborhood environment. Intuitively, remote sensing techniques would appear to have much greater applicability to the latter.

11.3.2. *Structural Quality of Individual Units*. National data on individual residential units are gathered by the United States Burea of Census through its decennial *Census of Housing* and by the American Public Health Association on a more or less continuous basis. Since 1973, the Census Burea has also been publishing an *Annual Housing Survey* based upon a sample of approximately 75,000 housing units drawn from 15 SMSA's. The data acquired in this survey include such items as tenure and race of occupant, number and type of rooms, condition of kitchen and bathroom facilities, type of sewage disposal, and condition of structure. In general, remote sensing techniques can be of little help in providing information of this type. However, the previous discussion on population estimation techniques revealed that information could be provided on the estimated number of dwelling units, the number of floors, architectural style, number of chimneys, presence of porches, fire escapes, carports and garages, and the amount of yardspace and parking space. While these data cannot necessarily be thought of as a substitute for *Survey* data, they can certainly supplement them.

11.3.3. *Neighborhood/Environmental Quality*. Remote sensing would appear to be an especially effective technique for providing data on residential quality, particularly at the neighborhood and community levels. Research in this area has generally entailed one of two approaches. The first approach has been to compare those data on environmental quality acquired by remote sensing with those acquired by more convential ground-survey methods. The second has been to test the relationship between housing density and a variety of socioeconomic factors. Both approaches have demonstrated some success.

In one example of the comparative approach, Marble and Horton examined housing quality data gathered by the Los Angeles County Health Department and concluded that at least 21 of the variables utilized by the Department were potentially measureable by remote sensing. Of these, seven were shown to be sufficient for discriminating between various housing quality classes: on-street parking; loading and parking hazards; street width; hazards from traffic; refuse; street grade; and access to buildings. When these variables were analyzed on aerial photos, it was found that the seven could be reduced to four (street width, on-street parking, street grade, and hazards from traffic) with only a slight reduction (82 percent to 78 percent) in the accuracy of assigning blocks to housing quality classes [148].

Howard, *et al.* employed a similar approach comparing residential quality data extracted from color-infrared imagery (1 : 6,000) with those gathered by the

American Public Health Association [149]. Of the thirty-seven environmental criteria used in determining residential quality, thirty-one could be acquired by remote sensing. Several of these criteria were applied to a sample of Denver neighborhoods to determine residential quality. While the authors found the approach 'promising', they failed by their own admission to provide any cost comparisons between the remote sensing and APHA systems.

Concerning the relationship between housing density and socioeconomic conditions, Metivier and McCoy had observed that density of single-family housing was an effective surrogate for poverty [150]. In a follow-up study [151], the same authors employed simple correlation and regression analysis to examine the relationships between density of single-family housing and five socioeconomic variables: percent owner occupied, percent renter occupied, house value, average rent, and median family income. Tests of significance showed relationships between density and income, density and house value, and density and average rent to be significant at the 0.01 level. A similar study undertaken by Henderson and Utano corroborated these results. Density of single-family housing displayed a strong correlation with average rent, average house value, median family income, and average number of rooms [152].

The ability to evaluate housing quality remote sensing has important implications for the field of public health where recent studies have demonstrated an association between poverty and disease. Accordingly, Rush and Vernon employed low-level color photography (1:6,000 and 1:12,000) to acquire data on the density and quality of selected residential areas [153]. These data, as well as census data on average rent, housing value, percent renter occupied, and property value, were compared to a variety of health statistics to determine the degree of relationship. Although neither set of data were found to be superior, those acquired by remote sensing were found to be as useful as census data in determining health outcomes.

The little research that has taken place on the application of remote sensing techniques to the acquisition of housing data has focused primarily on single-family housing and on the United States. There has also been a failure to provide estimates of comparative costs of remote sensing methods versus the more conventional field-survey methods. Nevertheless, the results of the available research have been sufficiently encouraging to warrant their being brought to the attention of housing authorities, particularly those Third World authorities faced with an almost total absence of reliable housing data.

References

1. Department of the Army. *Remote Sensing Applications Guide*, Engineering Pamphlet 70-1-1, (Washington, D.C.), 1979.
2. Strahler, A.N. *Physical Geography* (John Wiley and Sons, Inc., New York), 1975.
3. Colwell, R. The visible portion of the spectrum. In: Lintz J. Jr. and D.S. Simonett (eds.), *Remote Sensing of Environment* (Addison-Wesley Publishing, Reading, MA.), 1976.
4. Eastman Kodak. *Kodak Data for Aerial Photography* (Rochester, N.Y.), 1971.
5. Wenderoth, S. and E. Yost. *Multispectral Photography for Earth Resources* (Long Island University, New York), 1974.
6. Heller, R.C. Imaging with photographic sensors. In: *Remote Sensing with Special Reference to Agriculture and Forestry* (National Academy of Sciences, Washington D.C.), 1970.
7. Gausman, H.W. Leaf reflectance of near-infrared. *Photogrammetric Engineering*, 40.2: 183–191, 1974.
8. Specht, M.R. IR and pan films. *Photogrammetric Engineering*, 36.3: 360–364, 1970.
9. Fritz, N. Optimum methods for using infrared-sensitive color films. *Photogrammetric Engineering*, 33.10: 1128–1138, 1967.
10. White, L.P. *Aerial Photography and Remote Sensing for Soil Survey* (Clarendon Press, Oxford), 1977.
11. Westfall, C.Z. *Map Drawing* (University of Maine, Orono), 1973.
12. Kilford, W.K. *Elementary Air Survey* (Pitman Publishing, London), 1970.
13. Paine, D.P. *Aerial Photography and Image Interpretation for Resource Management* (John Wiley and Sons, New York), 1981.
14. Rea, J.C. and M. Ashley. A camera system for small format aerial photography. In: F. Shahrokhi, *Remote Sensing of Earth Resources* (University of Tennessee Space Institute, Tullahoma), 1977.
15. Fleming, J. and R.G. Dixon. *Basic Guide to Small Format Hand-Held Oblique Aerial Photography* (Energy, Mines and Resources, Canada), 1981.
16. Watson, E.K. Applications of 35 mm aerial photography to ecological land survey. *Canadial Journal of Remote Sensing*, 9.1: 31–44, 1983.
17. Meyer, M., S. Opseth, N. Moody and L. Bergstrom. *Large-Area Forest Land Management Applications of Small Scale 35 mm Aerial Photography*, Research Report 82–3 (Remote Sensing Laboratory, University Of Minnesota, Minneapolis), 1982.
18. Kinnucan, P. Earth-scanning satellites lead resource hunt. *High Technology*, March/April, 1982.
19. Short, N.M. *The Landsat Tutorial Workbook: Basics of Satellite Remote Sensing* (NASA, Washington, D.C.), 1982.
20. Robinson, A., R. Sale and J. Morrison. *Elements of Cartography* (John Wiley and Sons, New York), 1978.
21. Shelton, R.L. and J.E. Estes. Remote sensing and geographic information systems: an unrealized potential. *Geo-Processing*, 1.4: 395–420, 1981.

22. Minnesota Land Management Information Center. *LMIC Overview 1982* (St. Paul, Minnesota).
23. Jackson, R.H. *Land Use in America* (John Wiley and Sons, New York), 1981.
24. Burley, T.M. Land use or land utilization? *Professional Geographer*, 13.6: 18–20, 1971.
25. Anderson, J.R. Land use classification schemes used in selected recent geographic applications of remote sensing. *Photogrammetric Engineering*, 37.4: 379, 1971.
26. Simpson, R.B., D.T. Lindgren, D.J. Ruoml and W. Goldstein. *Investigation of Land Use of Northern Megalopolis Using ERTS-1 Imagery*, NASA CR-2459 (NASA, Washington, D.C.), 1974.
27. Kleckner, R.L. A national program of land use and land cover mapping and data compilation. *Planning Future Land Uses* (ASA, CSSA, SSSA, Madison), 1981.
28. Baker, R.D., J.E. DeSteigver, D.E. Grant and M.J. Newton. Land-use/land cover mapping from aerial photographs. *Photogrammetric Engineering and Remote Sensing*, 45.5: 661–668, 1979.
29. Campbell, J.B. *Mapping the Land: Aerial Imagery for Land Use Information* (Association of American Geographers, Washington, D.C.), 1983.
30. Anderson, J.R., E.E. Hardy, J.T. Roach and R.E. Witmer. *A Land-Use and Land Cover Classification System for Use with Remote Sensor Data*, Professional Paper 964 (U.S. Geological Survey, Washington, D.C.), 1976.
31. Milazzo, V.A. *A Review and Evaluation of Alternatives For Updating Geological Survey Land use and Land Cover Maps*, Geological Survey Circular 826 (U.S. Geological Survey, Reston, Virginia), 1980.
32. McConnell, W.P. and W. Neidzwiedz. *Remote Sensing 20 Years of Change in Worcester County, Massachusetts, 1951–1971*, Research Bulletin 625 (University of Massachusetts), 1974.
33. Lindgren, D.T., R.B. Simpson and W. Goldstein. *Land Use Change Detection in the Boston and New Haven Areas* (Dartmouth College Project in Remote Sensing, Hanover, New Hampshire), 1974.
34. Jensen, J.R. Urban change detection mapping using Landsat digital data. *The American Cartographer*, 8.2: 127–147, 1981.
35. Stow, D.A., L.R. Tinney and J.E. Estes. Deriving land use/land cover change statistics from Landsat: a study of prime agricultural lands. *Proceedings*, 14th International Symposium on Remote Sensing of Environment, San Jose, Costa Rica (The University of Michigan, Ann Arbor), 1227–1237, 1980.
36. Shepard, J.R. A concept of change detection. *Photogrammetric Engineering*, 30.4: 648–651, 1964.
37. Stauffer, M.L. and R.L. McKinney. *Landsat Image Differencing as an Automated Land Cover Change Detection Technique*, NAS 5-24350 (General Electric Company, Beltsville, Maryland), 1978.
38. Robinson, J.W. *A Critical Review of the Change Detection and Urban Classification Literature*, Technical Memorandum 79/6235 (Computer Sciences Corporation, Silver Spring, Maryland), 1979.
39. Jensen, J.R., J.E. Estes and L.R. Tinney. High-altitude versus Landsat imagery for digital crop identification. *Photogrammetric Engineering and Remote Sensing*, 44.6: 723–733, 1978.
40. Barlowe, R. *Land Resource Economics: The Economics of Real Estate* (Prentice-Hall, Englewood Cliffs, N.J.), 1978.
41. Long L. and D. DeAre. The slowing of urbanization in the U.S. *Scientific American*, 1.249: 33–41, 1983.
42. National Agricultural Lands Study. *Final Report* (Washington, D.C.), 1981.
43. Fischel, W.A. The urbanization of agricultural land: a review of the national agricultural lands study. *Land Economics*, 52.2: 236–259, 1982.
44. Soil Conservation Service. *Potential Cropland Study*, Statistical Bulletin 578 (U.S. Department of Agriculture, Washington, D.C.), 1977.

45. Vermont Department of Agriculture. *Mapping Farmland in Vermont: A Methodology* (Montpelier, Vermont), 1983.
46. Wildman, W.E. Detection and management of soil, irrigation, and drainage problems. In: Johannsen, C.J. and J.L. Sanders, *Remote Sensing for Resource Management* (Soil Conservation Society of America, Ankeny, Iowa), 1982.
47. American Society of Photogrammetry. *Manual of Remote Sensing*, Vol. II (ASP, Falls Church, Virginia), 1975.
48. Manzer, F.E. and G.R. Cooper. *Aerial Photographic Methods of Potato Disease Detection*, Bulletin 646 (University of Maine, Orono), 1967.
49. Myers, V.I. Soil, water, and plant relations. In: *Remote Sensing with Special Reference to Agriculture and Forestry* (National Academy of Sciences, Washington, D.C.), 1970.
50. White, J.G. NOW – infrared can spot your crop problems. *Farm Journal*, 12: 18–19, 1976.
51. American Society of Photogrammetry. *Manual of Remote Sensing*, Vol. II (ASP, Falls Church, Virginia), 1983.
52. DiPaolo, W.D. and L.B. Hall. Landsat data for soils investigations on federal lands. In: Johannsen, C.J. and J.L. Sanders, *Remote Sensing for Resource Management* (Soil Conservation Society of America, Ankeny, Iowa), 1982.
53. Satterwhite, M., W. Rice and J. Shipman. Using landform and vegetative factors to improve the interpretation of Landsat imagery. *Photogrammetric Engineering and Remote Sensing*, 50.1: 83–91, 1984.
54. Zonneveld, I.S. Land information, ecology and development. *ITC Journal*, 4: 475–498, 1979.
55. Spiers, B. A vegetation survey of semi-natural grazing lands (dehesas) near Merida, S.W. Spain. *ITC Journal*, 4: 649–679, 1978.
56. Colwell, R.N. Applications of remote sensing in agriculture and forestry. In: *Remote Sensing with Special Reference to Agriculture and Forestry* (National Academy of Sciences, Washington, D.C.), 1970.
57. Deuell, R.L. and T.M. Lillesand. An aerial photographic procedure for estimating recreational boating use on inland lakes. *Photogrammetric Engineering and Remote Sensing*, 11.48: 1713–1717, 1982.
58. Green, L.R., J.K. Olson, W.G. Hart and M.R. Davis. Aerial photographic detection of imported fire ant mounds. *Photogrammetric Engineering and Remote Sensing*, 43.8: 1051–1057, 1977.
59. Watson, T.C. *Aerial Photos and Pocket Gopher Populations*, M.Sc. Thesis (Colorado State University, Ft. Collins), 1973.
60. Hielkema, J. Desert locust habitat monitoring with satellite remote sensing. *ITC Journal*, 4: 387–417, 1981.
61. McCulloch, L. and D.M. Hunter. Identification and monitoring of Australian plague locust from Landsat. *Remote Sensing of Environment*, 13.2: 95–102, 1983.
62. Deshler, W. An examination of the extent of fire in the grassland and savanna of Africa along the southern side of the Sahara. *Proceedings*, Ninth International Symposium on Remote Sensing of Environment, Vol. I (The University of Michigan, Ann Arbor), 1974.
63. Strandberg, C.H. and C. Lukerman. *Development of Livestock and Crop Survey Techniques*, Report No. 118 (Itek Corporation, Palo Alto, California), 1983.
64. Chang, S.J. and J. Buongiorno. A programming model for multiple use forestry. *Journal of Environmental Management*, 13: 45–58, 1981.
65. Sayn-Wittgenstein, L. *Recognition of Tree Species on Aerial Photographs*, Information Report FMR-X-118 (Forest Management Institute, Ottawa, Ontario), 1978.
66. Dodge, A.G. Jr. and E.S. Bryant. Forest type mapping with satellite data. *Journal of Forestry*, 8.74: 526–531, 1976.
67. Middleton, E.M., B.G. Bly and J.A. Copony. *Virginia Forestry Project: A Landsat Survey of James City County*, ERRSAC Project Report 80-1 (Goddard Space Flight

Center, Greenbelt, Maryland), 1980.
68. Smit, G.S. Use of Landsat imagery for forest management. *ITC Journal*, 3: 563–575, 1980.
69. Stellingwerf, D.A. Orthophoto maps and/or Landsat Computer print-outs for forestry: aspects of a quantification problem. *ITC Journal*, 4: 499–517, 1979.
70. Avery, T.E. *Foresters Guide to Aerial Photo Interpretation*, Agricultural Handbook 308 (U.S. Department of Agriculture, Washington, D.C.), 1966.
71. Remeijn, J.M. Forest road planning from aerial photographs. *ITC Journal*, 3: 429–443, 1978.
72. Rasmussen, W.O., R.N. Weisz, P.F. Ffolliot and D.R. Carder. Planning for forest roads – a computer-assisted procedure for selection of alternative corridors. *Journal of Environmental Management*, 11: 93–104, 1980.
73. Hafker, W.R. and W.R. Philipson. Landsat detection of hardwood forest clearcuts. *Photogrammetric Engineering and Remote Sensing*, 5.48: 779–780, 1982.
74. Murtha, P.A. Some air-photo scale effects on Douglas-Fir damage type interpretation. *Photogrammetric Engineering and Remote Sensing*, 49.3: 327–335, 1983.
75. Ciesla, W.M. Color vs. color-infrared photos for forest insect surveys. *Proceedings*, 6th Biennial Workshop, Aerial Color Photography in Plant Science and Related Fields (Colorado State University, Fort Collins), 1977.
76. McCarthy, J., C.E. Olson Jr. and J.A. Witter. Evaluation of spruce-fir forests using small-format photographs. *Photogrammetric Engineering and Remote Sensing*, 48.5: 771–778, 1982.
77. Ashley, M.D., J. Rea and L. Wright. Spruce budworm damage evaluations using aerial photography. *Photogrammetric Engineering and Remote Sensing*, 42.10: 1265–1272, 1976.
78. Ciesla, W.M., D.D. Bennett and J.A. Caylor. Mapping effectiveness of insecticide treatments against pandora moth with color-IR photos. *Photogrammetric Engineering and Remote Sensing*, 50.1: 73–79, 1984.
79. Roundtree, R.A. and R.A. Sanders. *The Urban Forest Resource*, Report 13 (Dept. of Environmental Conservation, Albany, New York), 1981.
80. Lefor, M.W. and W.C. Kennard. *Inland Wetland Definitions*, Report 28 (University of Connecticut, Storrs), 1977.
81. Zetka, E.F. Coastal zone management information needs: potential Landsat applications. In: Johannsen, C.J. and J.L. Sanders, *Remote Sensing for Resource Management* (Soil Conservation Society of America, Ankeny, Iowa), 1982.
82. Carter, V. Applications of remote sensing to wetlands. In: Johannsen, C.J. and J.L. Sanders, *Remote Sensing for Resource Management* (Soil Conservation Society of America, Ankeny, Iowa), 1982.
83. Klemas, V., F.C. Diaber, D. Bartlett, O.W. Crichton and A.O. Fornes. Inventory of Delaware's wetlands. *Photogrammetric Engineering*, 40.4: 433–439, 1974.
84. Sharik, T.L., R.A. Meand, R.A. Eastmond and K.D. Hough. Use of color infrared photography in the delineation of wetlands vegetation in a Texas Gulf Coast watershed. *Canadian Journal of Remote Sensing*, 6.2: 73–85, 1980.
85. Seher, J.S. and P.T. Tueller. Color aerial photos for marshland. *Photogrammetric Engineering*, 39.5: 489–499, 1973.
86. Civco, D.L., W.C. Kennard and M.W. Lefor. A technique for evaluating inland wetland photointerpretation: the cell analytical method (CAM). *Photogrammetric Engineering and Remote Sensing*, 44.8: 1045–1052, 1978.
87. Grimes, B.H. and J.C.E. Hubbard. A comparison of film type and the importance of season for interpretation of coastal marshland vegetation. *Photogrammetric Record*, 7.38: 213–222, 1971.
88. Brown, W.W. Wetland mapping in New Jersey and New York. *Photogrammetric Engineering and Remote Sensing*, 44.3: 303–314, 1978.
89. Austin, A. and R. Adams. Aerial color and color infrared survey of marine plant resources. *Photogrammetric Engineering and Remote Sensing*, 44.4: 469–480, 1978.

90. Gammon, P.T. and V. Carter. Vegetative mapping with seasonal color infrared photographs. *Photogrammetric Engineering and Remote Sensing*, 45.1: 87–97, 1979.
91. Burgoyne, G.E. and L.G. Visser. Wildlife habitat evaluation demonstration project. *Eastern Regional Remote Sensing Applications Conference*, NASA Conference Publication CP 2173 (NASA, Washington, D.C.), 1981.
92. Mroczynski, R.P. and D. Eisenhauer. Aerial surveys for pheasant habitat. In: Johannsen, C.J. and J.L. Sanders, *Remote Sensing for Resource Management* (Soil Conservation Society of America, Ankeny, Iowa), 1982.
93. Mead, R.A., T.L. Sharik, S.P. Prisley and J.J. Heinen. A computerized spatial analysis system for assessing wildlife habitat from vegetation maps. *Canadian Journal of Remote Sensing*, 1.7: 34–40, 1981.
94. Lyon, J.G. Landsat-derived land-cover classifications for locating potential kestrel nesting habitat. *Photogrammetric Engineering and Remote Sensing*, 2.42: 245–250, 1983.
95. Wiersema, G. Ibex habitat analysis using Landsat imagery. *ITC Journal*, 2: 139–147, 1983.
96. Graves, H.G., E.D. Bellis and W.M. Knuth. Censusing Whitetailed Deer by airborne thermal infrared imagery. *Journal of Wildlife Management*, 3.36: 875–884, 1972.
97. Wyatt, C.L., M.M. Trivedi and D.R. Anderson. Statistical evaluation of remotely sensed thermal data for deer census. *Journal of Wildlife Management*, 2.44: 397–401, 1980.
98. Trivedi, M.M., C.L. Wyatt and D.R. Anderson. A multi-spectral approach to remote detection of deer. *Photogrammetric Engineering and Remote Sensing*, 12.48: 1879–1889, 1982.
99. Hampton, J.N. Aerial reconnaissance for archaeology: uses of photographic evidence. *Photogrammetric Record*, 9.50: 265–272, 1977.
100. Deuel, L. *Flights into Yesterday* (St. Martin's Press, New York), 1969.
101. Strandberg, C.H. Archaeological findings with color aerial photography. Convention Paper 199 (American Society of Photogrammetry, Falls Church, Virginia), 1967.
102. Martin, A.M. Archaeological sites – soils and climate. *Photogrammetric Engineering*, 4.37: 353–358, 1971.
103. Miller, W.F. Remote sensing techniques in historical site discrimination. *Remote Sensing of Environment*, 11: 463–471, 1981.
104. Ashley, M.D., D. Sange, H. Borns and L. Wright. Archaeological studies of forested sites using aerial photography. In: Shahrokhi, F., *Remote Sensing of Earth Resources* (University of Tennessee Space Institute, Tullahoma, Tennessee), 1977.
105. Hampton, J.N. An experiment in multispectral air photography for archaeological research. *Photogrammetric Record*, 8.43: 37–64, 1974.
106. Avery, T.E. and T.R. Lyons. *Remote Sensing: Aerial and Terrestrial Photography for Archeologists* (National Park Service, Washington, D.C.), 1981.
107. Johnson, G.W. and E.W. Kase. Mapping an ancient trade route with balloon photography. *Photogrammetric Engineering and Remote Sensing,* 43.12: 1489–1493, 1977.
108. Way, D.S. *Terrain Analysis: A Guide to Site Selection Using Aerial Photographic Interpretation* (Dowden, Hutchinson and Ross, Stroudsburg, PA.), 1971.
109. Bisset, R. Methods for environmental impact analysis: recent trends and future prospects. *Journal of Environmental Management,* 11: 27–43, 1980.
110. Hollick, M. Environmental impact assessment as a planning tool. *Journal of Environmental Management,* 12: 79–90, 1981.
111. Wobber, F.J. and K.R. Martin. Siting of nuclear power plants using remote sensor data. *Functional Photography,* 12: 22–26, 1977.
112. Hansen, K.L. Impact of demographic siting criteria and environmental suitability on land availability for nuclear reactor siting. *Proceedings,* National Conference on Energy Resource Management, Vol. II, NASA Conference Publication 2261 (Baltimore, MD.), 1982.

113. Barr, D.J. and M.D. Hensey. Industrial site study with remote sensing. *Photogrammetric Engineering,* 40.1: 79–86, 1974.
114. Wray, J.R. Potential role of land use and land cover information in powerplant siting: example of Three Mile Island. *Proceedings,* National Conference on Energy Resource Management, Vol. II, NASA Conference Publication 2261 (Baltimore, MD.), 1982.
115. Garfalo, D. and F.J. Wobber. Solid waste with remote sensing. *Photogrammetric Engineering*, 40.1: 45–60, 1974.
116. Turner, A.K. A decade of experience in computer route selection. *Photogrammetric Engineering and Remote Sensing*, 44.12: 1561–1576, 1978.
117. Hamilton, L.S. and D.S. Locate. *Using Air Photo Interpretation and a Socio-Ecological Reconnaissance in the Highway Route Selection Process* (Cornell University, Ithaca, New York), 1970.
118. Klunder, H. and R.B. Arend. Airphoto interpretation and the selection of a powerline right-of-way in Vermont. *Proceedings*, National Conference on Energy Resource Management, Vol. II, NASA Conference Publication 2261 (Baltimore, Maryland), 1982.
119. Grumstrup, P.D., M.P. Meyer, R.J. Gustafson and E.R. Hendrickson. Aerial photographic assessment of transmission line structure impact on agricultural crop production. *Photogrammetric Engineering and Remote Sensing*, 48.8: 1313–1317, 1982.
120. Branch, M.C. *City Planning and Aerial Information* (Harvard University Press, Cambridge, Massachusetts), 1971.
121. Quinn, A.O. Admissibility in court of photogrammetric products. *Photogrammetric Engineering and Remote Sensing*, 2.45: 167–170, 1979.
122. Earl Loesch, et. al. v. The United States, U.S. Court of Claims (Ohio), Opinion, October 31, 1979.
123. Lins, H.F. Jr. Some legal considerations in remote sensing. *Photogrammetric Engineering and Remote Sensing*, 6.45: 741–748, 1979.
124. Christian Science *Monitor*. Overflights irk Dow Chemical, March 26, 1980, p. 14.
125. Clemens, E. and L. Warnick. Remote sensing applications to the Pennsylvania abandoned mine reclamation program. *Proceedings*, National Conference on Energy Resource Management, Vol. II, NASA Conference Publications 2261 (Baltimore, Maryland), 1982.
126. Mroczynski, R.P. and R.A. Weismiller. Aerial photography: a tool for strip mine reclamation. In: Johannsen, C.J. and J.L. Sanders, *Remote Sensing for Resource Management* (Soil Conservation Society of America, Ankeny, Iowa), 1982.
127. Spisz, E.W. and J.T. Dooley. *Assessment of Satellite and Aircraft Multispectral Scanner Data for Strip-Mine monitoring*, NASA TM-79268 (Lewis Research Center, Cleveland, Ohio), 1979.
128. Davis, A.L., H.L. Bleomer and J.O. Brumfield. A temporal approach to monitor surface mine reclamation progress via Landsat. *Proceedings*, National Conference on Energy Resource Management, Vol. II, NASA Conference Publication 2261 (Baltimore, Maryland), 1982.
129. Adeniyi, P.O. An aerial photographic method for estimating urban population. *Photogrammetric Engineering and Remote Sensing*, 48.4: 545–560, 1983.
130. Henderson, F.M. Housing and population analysis. In: Ford, K. *Remote Sensing for Planners* (Center for Urban Policy Research, Rutgers, The State University of New Jersey), 1979.
131. Starsinic, D.E. and M. Zitter. Accuracy of housing unit method in preparing population estimates for cities. *Demography*, 5: 475–484, 1969.
132. Green, N.E. Aerial photographic analysis of residential neighborhoods: an evaluation of data accuracy. *Social Forces*, 35: 142–147, 1956.
133. Hadfield, S.A. *Evaluation of Land Use and Dwelling Unit Data Derived from Aerial Photography* (Chicago Area Transportation Study, Chicago, Illinois), 1963.

134. Binsell, R. *Dwelling Unit Estimation from Aerial Photography* (Department of Geography, Northwestern University, Evanston, Illinois), 1967.
135. Lindgren, D.T. Dwelling unit estimation with color infrared photos. *Photogrammetric Engineering*, 373–377, 1971.
136. Clayton, C. and J.E. Estes. Image analysis as a check on census enumeration accuracy. *Photogrammetric Engineering and Remote Sensing*, 46.6: 757–764, 1980.
137. Allan, J.A. and T. Alemayehu. Rural population estimates from air photographs: an example from Wolamo, Ethiopia. *ITC Journal*, 1: 85–100, 1975.
138. Lo, C.P. and F.F. Chan. Rural population estimation from aerial photographs. *Photogrammetric Engineering and Remote Sensing*, 46.3: 337–345, 1980.
139. Eyre, L.A., B. Adolphus and M. Amiel. Census analysis and population studies. *Photogrammetric Engineering*, 36.5: 460–466, 1970.
140. Collins, W.G. and A.H.A. El-Beik. Population census with aid of aerial photographs: an experiment in the city of Leeds. *Photogrammetric Record*, 7.37: 16–26, 1971.
141. Kraus, S.P., L.W. Senger and J.M. Ryerson. Estimating population from photographically determined residential land use types. *Remote Sensing of Environment*, 3: 35–42, 1974.
142. Polle, V.F.L. The size of net residential areas on aerial photographs. *ITC Journal*, 3: 544–561, 1980.
143. Nordbeck, S. *The Law of Allometric Growth*, Discussion Paper 7 (Department of Geography, University of Michigan, Ann Arbor), 1965.
144. Holz, R.K., D.K. Huff and R.C. Mayfield. Urban spatial structure based on remote sensing imagery. *Proceedings*, Sixth International Symposium on Remote Sensing of Environment, Vol. II (University of Michigan, Ann Arbor), 1969.
145. Ogrosky, C.B. Population estimates from satellite imagery. *Photogrammetric Engineering and Remote Sensing*, 41.6: 707–712, 1975.
146. Lo, C.P. and R. Welch. Chinese urban population estimates. *Annals*, Association of American Geographers, 67.2: 246–253, 1977.
147. Anderson, D.E. and P.N. Anderson. Population estimates by humans and machines. *Photogrammetric Engineering*, 39.2: 147–154, 1973.
148. Marble, D.F. and F.E. Horton. Extraction of urban data from high and low resolution images. *Proceedings*, Sixth International Symposium on Remote Sensing of Environment, Vol. II (University of Michigan, Ann Arbor), 1969.
149. Howard, W.A., L.C. Harold, L.B. Driscoll and L.R. LaRerrier. *Residential Environmental Quality in Denver Utilizing Remote Sensing Techniques*, Publications in Geography, 74–1 (University of Denver, Denver), 1974.
150. Metivier, E.D. and R.M. McCoy. Mapping urban poverty housing from aerial photographs. *Proceedings*, Seventh International Symposium on Remote Sensing of Environment, Vol. II (University of Michigan, Ann Arbor), 1971.
151. McCoy, R.M. and E.D. Metivier. House density versus socio-economic conditions. *Photogrammetric Engineering*, 39.1: 43–47, 1973.
152. Henderson, F.M. and J.J. Utano. Assessing general urban socio-economic conditions with conventional air photography. *Photogrammetria*, 31: 81–89, 1975.
153. Rush, M. and S. Vernon. Remote sensing and urban public health. *Photogrammetric Engineering and Remote Sensing*, 41.9: 1149–1155, 1975.

Index